THE COVID-19 PANDEMIC

THE COVID-19 PANDEMIC
A Global High-Tech Challenge at the Interface of Science, Politics, and Illusions

KLAUS ROSE
klausrose Consulting, Pediatric Drug Development and More, Riehen, Switzerland

Academic Press is an imprint of Elsevier
125 London Wall, London EC2Y 5AS, United Kingdom
525 B Street, Suite 1650, San Diego, CA 92101, United States
50 Hampshire Street, 5th Floor, Cambridge, MA 02139, United States
The Boulevard, Langford Lane, Kidlington, Oxford OX5 1GB, United Kingdom

Copyright © 2022 Elsevier Inc. All rights reserved.

No part of this publication may be reproduced or transmitted in any form or by any means, electronic or mechanical, including photocopying, recording, or any information storage and retrieval system, without permission in writing from the publisher. Details on how to seek permission, further information about the Publisher's permissions policies and our arrangements with organizations such as the Copyright Clearance Center and the Copyright Licensing Agency, can be found at our website: www.elsevier.com/permissions.

This book and the individual contributions contained in it are protected under copyright by the Publisher (other than as may be noted herein).

Notices
Knowledge and best practice in this field are constantly changing. As new research and experience broaden our understanding, changes in research methods, professional practices, or medical treatment may become necessary.

Practitioners and researchers must always rely on their own experience and knowledge in evaluating and using any information, methods, compounds, or experiments described herein. In using such information or methods they should be mindful of their own safety and the safety of others, including parties for whom they have a professional responsibility.

To the fullest extent of the law, neither the Publisher nor the authors, contributors, or editors, assume any liability for any injury and/or damage to persons or property as a matter of products liability, negligence or otherwise, or from any use or operation of any methods, products, instructions, or ideas contained in the material herein.

Library of Congress Cataloging-in-Publication Data
A catalog record for this book is available from the Library of Congress

British Library Cataloguing-in-Publication Data
A catalogue record for this book is available from the British Library

ISBN 978-0-323-99149-0

For information on all Academic Press publications
visit our website at https://www.elsevier.com/books-and-journals

Publisher: Stacy Masucci
Acquisitions Editor: Elizabeth Brown
Editorial Project Manager: Franchezca A. Cabural
Production Project Manager: Kiruthika Govindaraju
Cover Designer: Miles Hitchen

Typeset by STRAIVE, India

Dedication

This book is dedicated to our daughter Rebecca, without whom this book would not have been possible. She was our first child, born in 1994 with Sturge–Weber syndrome, a complex disease for which only symptomatic treatment exists. For details of this disease, see www.sturge-weber.org. Rebecca was a moderately mentally and physically challenged child and suffered from epilepsy, among other challenges. She was always full of zest for life. Her epilepsy had become rather stable, and we were looking forward to see how she would grow further, but in June 2021 she had a nocturnal epileptic fit and passed away of suffocation. Living with a special child shapes the life of parents and siblings and gives you an insight into areas and aspects of life that would otherwise remain closed. Rebecca will always stay alive in our hearts.

Contents

Foreword	xi
Preface	xiii

1. Introduction

1. Introduction	1
References	13

2. We are no longer hunters and gatherers. Societies, states, values, and healthcare today

3. Development of drugs and vaccines

4. COVID-19—The disease

1	Introduction	35
2	DNA and RNA	36
3	Severe acute respiratory syndrome coronavirus 2 (SARS-CoV-2) and COVID-19 disease	41
4	Modern society and public health	43
5	Traditional avenues of prevention	45
6	Vaccines	47
7	Diagnostics	53
8	Treatment	53
9	COVID-19, children, and "children"	56
10	COVID-19 variants	60
11	Conclusions	61
References		62

5. Russian and Chinese vaccines

1	Russian vaccines	71
2	Chinese vaccines	73
3	Assessment	75
References		76

6. The European Union (EU) response to the COVID-19 pandemic

1	The original course of the pandemic in Europe	79
2	Details of the EU response	80
3	Future EU plans	81
4	A preliminary assessment of the EU response to the COVID-19 pandemic	83
References		83

7. COVID-19 vaccines global access (COVAX) and more

1	Introduction	85
2	Key international organizations	85
3	COVID-19 vaccines global access (COVAX)	87
4	The discussion about booster shots	87
5	Preliminary COVAX assessment	88
References		88

8. International healthcare structures and COVID-19

1	WHO basics and "Public Health Emergencies of International Concern" (PHEICs)	91
2	International Health regulations (IHR) and PHEICs	92
3	The WHO's life of its own	93
4	The WHO and progress in healthcare	96
5	The WHO in the COVID-19 pandemic	99
6	The independent panel for pandemic preparedness and response (IPPR) report	101
7	Assessment of the WHO IPPR recommendations	103
8	Conclusions	104
References		105

9. Low-tech and high-tech challenges. Accidents and disasters. Technical and scientific progress and its perception by science and the public

1	Introduction	109
2	Groupthink	110
3	US space shuttle disasters	111
4	Boeing 737 MAX crashes	113
5	Nuclear plant meltdown in Fukushima, Japan, after a tsunami	114
6	Lead poisoning	115
7	Love Canal/Blackcreek village	119
8	Mercury poisoning	120
9	Bhopal	121
10	Tobacco smoking	121
11	BSE and Jacob-Creutzfeld-disease	123
12	Intermediate summary	125
13	Discussion and conclusions	126
References		129

10. Basic research, applied research, and the real world

11. Conflicts of interest and the self-picture of medicine and scientists

1 Introduction	141
2 Science is part of society	145
3 The special edition on 150 years "Nature" in 2019	147
4 The "Nature" coverage of the European Union (EU) science budget deliberations in 2019	149
5 The term "medical-industrial complex" and its ramifications	151
6 Conclusions	153
References	154

12. Vaccination hesitancy

1 Introduction	159
2 Discussion	166
3 Additional dimensions	167
4 Conclusions	168
References	169

13. Social inequality, developing countries, and COVID-19

1 Introduction	171
2 Social framework and the responsibility for one's fate and health	174
3 Social inequality: The sociology approach	176
4 Ideologies and politics that promise(d) to abolish poverty	177
5 Developing countries and the COVID-10 pandemic	181
6 The misconception of a weakening of intellectual property as a way out of the pandemic	183
7 Conclusions	186
References	186

14. Politics, illusions, websites, and the real world

1 Humanity and communication	189
2 Fairy tales, oral tradition, the worlds of radio and television, and the internet	191
3 Politics, websites, and the real world	192
4 The COVID-19 pandemic and the internet	193
5 Conclusions	195
References	195

15. Global warming, Armageddon warnings, and the COVID-19 pandemic

1 Climate change and global warming: The basics	197
2 The predicted effect of climate change on humans	201
3 Global warming and the COVID-19 pandemic	201

4	The mixing of geoscience, social, and medical challenges	205
5	Scientific warnings in the past	205
6	The privileges of youth	208
7	COVID-19 is a high-tech challenge	208
References		209

16. China and Russia are giants on feet of clay

17. Conclusions and outlook

1	Conclusions	226
References		228

Index   **231**

Foreword

COVID-19 has altered the world forever. Unlike past (perhaps even more dramatic) pandemics, this is the first pandemic to occur in the internet era and our current reality of universal and continuous communications and news. Every minute of every day we were bombarded with statistics and images of what was transpiring around the world. The impact of the rising death count; the reality of the isolation, despair, and suffering of those expiring without being surrounded by loved ones; the impact on our first responders; and the view of the locked down and masked world will have consequences for decades, particularly on the younger members of our population. Images of major streets in major cities empty were haunting. The mandated lockdown resulted in world uneasiness that exploded and was manifested in overt lawlessness, racism, looting, rioting, and hypocritical behavior of those in charge.

How did the governments of the world and the pharmaceutical industry respond? There are clearly some amazing success stories that demonstrate the brilliance, work ethics, and dedication of scientists when afforded the correct resources. The appearance of effective and multiple vaccines in a matter of months is truly remarkable. No matter your political views, we can all celebrate these remarkable successes. However, there were also some strategic errors made. Klaus Rose has laid out the historical unfolding of this pandemic with an eye to the role that government and pharmaceutical companies have played in delaying potentially life-saving care to our youth. Treatments and vaccines have been withheld from those who are minors and at the time of writing of this foreword are still being withheld from those 12 years of age and younger.

I am both a parent and grandparent. My eldest grandson is now 14 years old. Last year, he towered over me and weighed more than 75 pounds more than me. Despite this reality, he was not vaccinated until very recently, while I was vaccinated 9 months ago. He lives in Texas; he was attending school and being exposed to other children, teachers, and family members, yet he was not given the protection (for himself and others) that a vaccine could have afforded him until approximately 2 months ago—7 months after his parents and grandparents were vaccinated. At present, his three younger brothers remain unvaccinated despite their attending school and sporting events and social events.

Our children and grandchildren are the world's most precious resource. It is time that the bureaucrats in government and the pharmaceutical industry treat them as such. Except for dosing based upon weight and special attention to newborns and premature babies, children should be given the same access to appropriate treatments and vaccines as adults. If a medication works on a 20-year-old, it most assuredly will work on a 12-year-old who may be physiologically bigger than the 20-year-old despite being emotionally less mature.

Klaus Rose takes on the journey of why this discrepancy and ultimately unethical attitude arose at its inception and still persists. Although most rules were initiated in an attempt to protect our young, they have now been complicated by conflicts of interest that have allowed them to persevere inappropriately. In my opinion, this outstanding book will serve as an essential reading material for all medical, law, and pharmacology students. In addition, all those involved in government oversight vis-à-vis the medical and pharmacology industry should also be mandated to read this book prior to taking on these administrative roles. I hope you will agree with Dr. Rose that the time has come to change how we treat our children!

Jane M. Grant-Kels[a,b,c,d]
[a]Dermatology, Pathology and Pediatrics, University of CT Health Center Dermatology Department, Farmington, CT, United States
[b]Department of Dermatology, University of CT Health Center Dermatology Department, Farmington, CT, United States
[c]Director of the Cutaneous Oncology Center and Melanoma Program, University of CT Health Center Dermatology Department, Farmington, CT, United States
[d]Dermatology Residency and Dermatopathology Lab, University of CT Health Center Dermatology Department, Farmington, CT, United States

Preface

There was no overall political leadership in the COVID-19. At least, the pharmaceutical and life science industry has paved the way out but is not enthusiastically praised for it. The United States was partially paralyzed. Russia and China hoped to gain diplomatic profile worldwide with homemade vaccines, but the efficacy of their vaccines is limited and trust in the quality of their clinical data is rightly low. The European Union (EU) did not order enough vaccines in time and then sued a large manufacturer for delivery delays. It would be too easy to blame the mistakes of governments and leaders on stupidity. Misjudgments abounded, idiocy exists, but blaming idiocy explains little on its own. Many scientists, politicians, the leaders of the EU and of the World Health Organization (WHO) have flawed illusions. They overestimate clinical medicine and are too influenced by the demonization of the pharmaceutical industry, which is on the defensive since decades. Accusing the industry of alleged greed has become a business of its own. Industry must be profit-oriented and competitive. That is the way how business works. The WHO and many scientists believe they are leaders because what they say is international and therefore right, which is another illusion. Russia and China with their nationalist dreams could not provide truly safe and effective vaccines when there were not yet enough available worldwide.

The origin of this book lies in the rather special area of pediatric drug development, in which I worked for a good 20 years. I began, like most people in their apprenticeship years, with great enthusiasm. As a medical doctor in the pharmaceutical industry, I thought I could contribute to better medicines for children. I was the global head of this area in two large international companies for together 10 years, followed by 10 more years as an independent consultant. Then something unexpected happened. I realized that "better medicines for children" do not really aim at improving children's health. That is what specialists and key representatives say, but not what they mean. It is an illusion that they try to maintain. There are strong conflicts of interest, just not as easily visible as those where specialists were bribed to lie openly or deceive cunningly, as it happened in the defense of leaded gasoline or tobacco smoking. Our world has become more complicated and complex. I identified conflicts of interest at the interface of clinical science, regulatory authorities, and industry. To put it more generally, I identified conflicts of interest at the interface of medicine and law. Thousands of clinical studies enroll worldwide minors, but most are medically pointless and many even harm. They are requested and done because the involved researchers advance their academic careers. Regulators display their concern for children in the court of public opinion, which translates into political capital, higher budgets, and careers. Industry functionaries participate.

In the response to the pandemic, conflicts of interest, blind spots, and illusions were the major obstacles to dealing with the pandemic. There is no world government. There was no real leadership. Many institutions claimed and claim such a leadership. I learned to distinguish between what people, institutions, and representatives of institutions say and what they mean. Why could 15-year-olds not be vaccinated against COVID-19 once the vaccines were known to work? Why do Institutional Review Boards/ethics committees allow "pediatric" vaccination studies in up to 17-year-olds, although 17-year-olds are already bodily mature? They should have been vaccinated with an approved mRNA-based vaccine as soon as possible. Young people were hardly affected at the beginning of the pandemic, but this is now changing with more infectious virus variants. Are we prepared for the next pandemic and a new infectious disease that kills young people first?

Many background details in this book are openly available on the Internet. However, you must have learned to distinguish the "infodemic" from relevant information. In the past, I would have needed a library, but online encyclopedias helped me remember things that I had already picked up somewhere in the past. For medical and scientific issues I used textbooks and up-to-date scientific publications listed in PubMed (https://pubmed.ncbi.nlm.nih.gov), where many, but not all, articles are open source.

There is the misconception in science that to penetrate deeper, you need computer-based processing of "big data." That may work sometimes, but not for the fundamental questions at the interface of medicine, law, and society. Without decades of life experience, my wife, our physically and intellectually challenged and our younger, healthy daughter, I would have remained unable to address the taboos that we need to unveil. This book carries you in a whirlwind through the light and dark sides of history and humanity, highlighting strengths and weaknesses. Hopefully, it will contribute to improving things. Enjoy reading the book.

CHAPTER 1

Introduction

1 Introduction

The COVID-19 pandemic has still a firm grip on the world, although in 2022 we see already the light at the end of the tunnel. Nevertheless, COVID-19 has killed millions of persons worldwide, still affects massively the economy, and has restricted our quality of life for a relevant period by closing cinemas, theaters, and other cultural offerings, sports events, and more. Shopping only with a face mask. Scared people. Working from home—those that kept a job. But many have lost their jobs. Depending on their country, they were helped, or not.

There are parallels to epidemics in the past and fundamental differences. COVID-19 would in the past probably not even have been regarded as a serious problem, let alone an epidemic. Only a minority of people who get infected become seriously ill, while epidemics like the plague, smallpox, cholera, or typhus killed people of all ages and wiped out a relevant part of the population. Epidemics in the past had their causes, but people did not know them at the time. And there was no effective treatment. Healers promised treatment. Some got rich, others died themselves. Towns tried to protect themselves by quarantining people or closing city gates. Mostly it was too late. There were no microscopes yet. Bacteria and viruses cannot be seen with the naked eye. Some measures made things even worse, such as killing all dogs and cats in Renaissance Italy, allowing rats and their fleas to reproduce more easily. Their fleas carried the bacterium that caused the plague.

Epidemics and pandemics have many different dimensions and perspectives. Today, everyone can watch what is happening on TV. On a personal level, individuals experience COVID-19 if they themselves, a life partner, a member of the family, or a friend falls ill. Most experience it through the restrictions of daily life. It is different for people who work in health care, in funeral homes, as journalists, in the police, the military, or other services and institutions that maintain public order. Yet another dimension is governments or local leaderships, which for more than a year now were continuously informed that there were no more free beds in intensive care, not enough ventilators, protective clothing, face masks, and that the healthcare system was on the verge of collapse. Some governments refuse(d) to accept even the most basic observations, others tried to hide facts and lied more or less openly. Virtually all governments and local leaders tend to gloss over their efforts and describe the situation as a series of successful and promising measures, triggered by their leadership.

COVID-19 is not characterized by piles of corpses on the street or people having retreated to their homes in whole quarters to die there. COVID-19 affects a highly developed civilization where for a century people are getting older, child mortality is declining, and more and more people live with conditions that previously were lethal, including insulin-dependent diabetes; cystic fibrosis; childhood leukemia; seniors with a broken hip joint who would be bedridden and would die without a joint replacement; and many more. Today, the value of human life regardless of age is considered sacrosanct. When we have an accident, a stroke, or get sick for any other reason, we expect to be transported to a hospital and receive effective treatment. The one factor that is really indicative of the likelihood of dying from COVID-19 is old age. Other factors are poor general condition, obesity, an impaired immune system, and other comorbidities.

What doctors, health professionals, politicians, and others can do about a new disease reflects what we know about the disease, how well we can diagnose it, what means of prevention and treatment we have, and to what extent we are prepared to use existing resources. At the pandemic's beginning, 15-year-olds could not be vaccinated with the Moderna vaccine, which had FDA emergency use authorization (EUA) from 18 years on only.[1,2] The Pfizer-BioNTech vaccine has now been FDA EUA from 12 years on, while for the first half of 2021 the lower age limit was 16 years.[3,4] Was the Moderna vaccine unsafe in 15-years-olds? No. Was the BioNTech vaccine unsafe in 11-year-olds in the first half of 2021? No. But the FDA and the European Medicines Agency (EMA) asked for separate "pediatric" data for these young persons, who mostly are bodily no longer children. Young women of 15 years can have children themselves, young men of 15 can father children, both can be athletes, taller and stronger than their parents. You do not have to have studied medicine to acknowledge that they are *bodily* already mature, although they are *legally* still minors. The age limit of COVID-19 vaccines is an issue of vaccine development and approval at the interface of medicine and law. We will encounter this question more often. Suffice it to say for now that in today's society drugs can only be used after a complicated approval process. We are *technically* able to develop good drugs and vaccines. The reasons why our world is not able to do against the COVID-19 pandemic everything that would *technically* be possible are complex. To understand them, we need to go back in history and understand the interaction between the various institutions, structures of power, and other social factors that play into the issues of maintaining public health.

Society, health care, and their interactions have changed profoundly during the past hundred years. Interventions are constantly further developed by a large machinery of research and development (R&D), allowing today keyhole repair of knees after sports injuries, surgical correction of congenital heart defects in infancy, cosmetic surgery, development of ever more effective drugs and vaccines, and more. R&D is at the interface of basic science, applied science, marketing, project management, and more. It uses the highest level of science available, but the primary aim of R&D is not publications in scientific journals. R&D is primarily carried out by commercial companies that have to survive in the harsh competitive environment of the market. It is not done by individual researchers in garage laboratories but is an organized and industrial process. Where in clinical research people are directly subjected to experimental treatment, the effectiveness of which has yet to be proven, R&D is tightly controlled by powerful regulatory authorities and other government institutions. Once R&D has delivered new drugs, technologies or devices, and these have been approved, their treatment is usually reimbursed. From then on, the medical community uses them hands-on, tries them out on a larger scale, discusses these further uses, and thus develops their use further.

In the COVID-19 pandemic, medical doctors have learned a lot since the first patients were hospitalized. Doctors relieve symptoms by antivirals, antiinflammatory medications, monoclonal antibodies, and more, as well as support breathing by oxygen. Initially, many patients were placed on ventilators, and it was feared that a shortage of ventilators would become the critical weak point in COVID-19 treatment. Later, ventilators were increasingly used as a last resort only. Efforts are made to prevent the excessive reaction of the immune system in patients, primarily the cytokine storm, which is often fatal once it has broken out.[5]

When President Trump was ill with COVID-19, he received eight drugs: dexamethasone; remdesivir; a monoclonal antibody; zinc, vitamin D, famotidine, melatonin, and aspirin.[6] The number of medical publications on COVID-19 is increasing exponentially, with over 200,000 PubMed-listed papers in January 2022.[7]

In the long term, viral infectious diseases can only be prevented by vaccination. The one exception is the human immunodeficiency virus (HIV), for which we still have no effective vaccines. However, drugs allow an almost normal life, provided they are taken correctly.[8] But that is another story. Vaccination is the most effective medical intervention in history. Together with clean water and sanitation, it has eliminated a large part of infectious diseases that once killed millions, specifically in childhood.[9] The smallpox vaccine came first. In 1796, it was shown that infection with the relatively mild cowpox conferred immunity against deadly smallpox.

The first vaccinations took a small amount of serum directly from the living cow and injected it into the vaccinated person, see Pictures 1 and 2, below, from 1895 and 1905, which illustrate why the term "vaccine" comes from "vacca," Latin for cow. In these pictures, the cow from which the inoculated fluid is taken is directly present in the room where the vaccinations took place. Smallpox became the first and until now only human disease to be eradicated. Smallpox vaccine is still being produced and stored as preparation for bioterrorism and biological warfare.[10] Eventually, the development of drugs and vaccines became a high-tech challenge. In the past, it took 10 years to develop a new vaccine. This has changed dramatically.

In both images, the cow is highlighted with a circle.

While the past US president underestimated the danger of the COVID-19 pandemic, refused to carry a mask, ridiculed himself in daily statements that showed how little he understood and cared, and in the end missed his re-election, his government also initiated "Operation Warp Speed" (OWS) in 2020, a public-private partnership to facilitate and

PICTURE 1 Vaccination 1895: French army recruits being vaccinated (Alfred Touchemolin).

PICTURE 2 Vaccination free of charge in the rooms of the French journal "Petit Journal," 1905.

accelerate the development, manufacturing, and distribution of COVID-19 vaccines, therapeutics, and diagnostics. The term "warp drive" comes from the science fiction series Star Trek. It stands for a (fictional) drive that enables faster-than-light travel by bending space–time. There was and is no historical predecessor for OWS, with the exception, perhaps, of the Manhattan Project, which developed the atomic bomb during World War II.[11] Usually, the development of drugs and vaccines is market-driven. Considering the dimension of the COVID-19 pandemic, OWS was a justified risk. And it worked. Effective vaccines were developed and FDA-approved in less than a year. The US government ordered early enough vaccines from all companies thought to have a reasonable chance to get a vaccine FDA-approved. The US government invested first about $ US 12 billion, later increased to about $18 billion, negligible expenditure compared to the overall costs of the COVID-19 pandemic so far. OWS included components of the Department of Health and Human Services (DHHS), the Centers for Disease Control and Prevention (CDC), the FDA, the National Institutes of Health (NIH), the Biomedical Advanced Research and Development Authority (BARDA); the Department of Defense; companies, and more. They also put an experienced industry chief executive officer (CEO) in the driver's seat.[12]

The parallel with the Manhattan Project is often drawn in the medical literature as an example of government intervention in a life-and-death situation.[13,14] A less known aspect of the Manhattan Project was that the same US government agency that oversaw the Manhattan Project, the Office of Scientific Research and Development, welcomed with open arms the English scientists who came to the US during WWII to promote the industrial production of penicillin. The US governmental agency opened the doors to the—then still—chemical industry. The result was the successful large-scale production of penicillin, which heralded a new era in medicine.[11]

Also the European Union (EU) supported the development of COVID-19 drugs and vaccines. The EU Commission offered the company CureVac financial support by backing a €75 million European Investment Bank loan. The European Investment Bank signed a €100 million financing agreement with immunotherapy company BioNTech.[15] The German government granted BioNTech €375 million for its COVID-19 vaccine development program.[16] Compared with the US approach, the EU approach was dimensions smaller. Furthermore, the funds were assigned to a multitude of different projects, and no capable leadership was established.[17]

During the first year of the COVID-19 pandemic, there was worldwide chaos, actionism, travel bans, finally curfews, school closings, lockdown, border closings, and more. Over time it became widely accepted that face masks protect against the spread of the SARS-CoV-2 virus to some extent. Now, more than a year after the pandemic began, we are also seeing the limitations of the conservative, defensive means. The real way out of the pandemic is through vaccination, a lesson we learned from dealing with other viral epidemics in the past, even though past viral epidemics hit completely different societies.

Today, we are in a new phase of the pandemic. Three vaccines have FDA Emergency Use Authorization (Pfizer-BioNTech, Moderna, Jannsen)[1–3]; the Pfizer-BioNTech vaccine has full FDA approval[18]; four have conditional approval by the European Medicines Agency (EMA) and by the United Kingdom (UK) Medicines and Healthcare products Regulatory Agency (MHRA) (Pfizer-BioNTech, Moderna, AstraZeneca, Jannsen),[19,20] one by the Russian authorities (in two formulations), and four by the Chinese authorities.[21] With OWS, the US made the biggest contribution. High-tech is always expensive in the beginning. In life-or-death situations, it is better not to turn over the penny twice, especially if much larger funds are wasted for other purposes.[17]

The UK was the first Western country to approve a COVID-19 vaccine, followed by the US FDA. The EMA, EU counterpart to the FDA, allowed itself an extended beauty sleep. It had to be kicked in the buttocks by health ministers until it reluctantly authorized the first vaccine. Why should it have hurried? It was only the lives of many people who had to die with every single day of later vaccine approval. Who would dare to challenge the sacred, orderly, drawn-out procedures of the EMA about such a triviality? Has the head of the EMA been fired? Has the chairperson of the EU commission resigned? Instead, there were accusations against AstraZeneca that had successfully developed one of the first vaccines. The usual political theater: blame somebody, anybody. The EU put export bans in place for vaccines, after it had hesitated for months to order enough, resulting in a disproportional workload for all vaccine developers to ship vaccines for ongoing clinical trials outside of the EU. Some months later, the EU sued AstraZeneca because too little vaccine was delivered too late. Vaccines are not supplied by God but must be physically produced on this earth. Most vaccines destined for the EU are produced in a Belgian plant, but there are more plants in other countries. The contract obliged AstraZeneca to make its best effort to meet the EU orders, which it did. AstraZeneca also acknowledged the option of supplying Europe from UK sites once the UK had sufficient supplies. The UK contract with AstraZeneca contained harsher clauses and seems to have been more thought out from the start.[22] In the meantime, the EU sued AstraZeneca a second time in a Belgian court, asking for compensation for the delayed delivery.[23] It will be a long time before the court decides. These lawsuits will certainly not accelerate the physical delivery of the vaccines. For a limited time, they may help the EU Leadership in its effort to save face. In the long run, such judicial activism will only show more clearly that the statesmanship of the EU leadership rather resembles a stage play in an amateur theater.

The one country that started vaccinations as soon as possible was Israel. Comparable to its military defense policy against hostile countries and organizations in its vicinity, Israel cannot afford to rest on the laurels of the past.[24-28] Obviously, the EU leadership can afford this luxury.

Did all countries start the vaccination as soon as possible, and do we acknowledge those players who made essential contributions? Are the delays in providing protective clothing, masks, ventilators, oxygen, and more the result of the stupidity of individuals? The stupidity of some or many individuals is annoying but does not explain much. On a tactical level, the US leadership made massive mistakes that were quite similar to those made worldwide in many other countries. But on the strategic level, which offers the real way out of the pandemic, the US has assumed global leadership with Operation Warp Speed, resulting in drastically accelerated vaccine development, fully employing the potential of modern biomedical industry and product development. To put it provocatively: the US is a strong country with a strong society that even survives the short-sighted tactical mistakes of an incompetent president. In the long term, the right strategy makes the difference, even if the many deaths caused by the shortsightedness of the former US president were horrible, unnecessary, and unfortunate. The key thesis of this book is that many mistakes that were made in tackling the COVID-19 pandemic were and are not based on the shortsightedness or stupidity of individuals (which would make changing course easy), but rather systematic mistakes that have crept into societal ideas about medicine and public health, and have found expression in the social power of various social institutions. We come to the crux of the matter. The topics in this book are very broad, which, however, is necessary to grasp the fundamental depth of the challenges the COVID-19 pandemic poses and to show us the next steps necessary to really move forward.

In the 60 years that have passed since the thalidomide catastrophe, the development of drugs and medical devices, medicine in general, and public health has made giant strides forward. Now we come up against limits set by the intellectual, social, and administrative framework in which technical progress takes place.

Three major movements have taken their course around the world since the thalidomide disaster. **(1)** The development of drugs and medical devices has been very successful. The pharmaceutical industry, today increasingly called the life science industry, became economically very successful. This has generated much criticism of alleged greed. The special circumstance of drug development is that the development costs are very high, especially because of the expensive clinical studies.[29] In contrast, the production costs, once drug efficacy has been proven and approved, are comparatively low. This has led to criticism over the decades, mainly looking at the relatively low production costs and inferring from this the alleged greed of the pharmaceutical industry.[30] Every company that wants to survive must be greedy to survive the ruthless competition of economic reality. To some degree, elements such as corporate social responsibility are possible,[31] but in the end, companies must survive. Of course, there have been thousands of attempts to make extra profits in illegal, criminal, or repulsive ways. Some or many of these attempts have even been successful. This has nothing to do with drugs or medical devices in themselves. Where there is a lot of money to be had, there are temptations. Open societies can deal with such temptations. Control mechanisms are not rocket science. **(2)** A concept has emerged in academic medicine under the constant barrage of the accusation of the greedy pharmaceutical industry, making the "medical-industrial complex" responsible for almost everything evil in the world, using it as a counter-symbol for an allegedly ethical, noble medicine that is concerned only with the patient, his health, and his well-being.[31,32] There are almost countless books that accuse the pharmaceutical industry of everything bad under the sky,[30–37] including putting it on the same level as organized crime.[38] **(3)** Assuming that industry is bad and motivated by greed had as a ramification the idea to reward pharmaceutical companies for "pediatric" studies, defining "children" as everybody younger than 17 years of age.[39] The initial age limit of 17 years shifted eventually to an internationally accepted age limit of 18 years.[40] The justification for these studies was the narrative of academic medical research and the Food and Drug Administration that **(a)** these studies were needed to improve drug treatment in children and **(b)** that for industry the additional costs for these studies would be prohibitive. Therefore, as an additional financial incentive 6 months patent extension was introduced in the US in 1997.[40,41] At least children should be protected from the pharmaceutical industry's greed. The concept of "pediatric drug development" sounds very convincing as long as we associate "children" with small, vulnerable human beings. But there is one big catch. Drugs can be dangerous if given in the wrong doses. This is especially true of babies, and even more so of premature babies. Drug toxicities had been observed in preterm newborns in the 1950s. But the narrative that from these observations expanded alleged toxicities from babies to all "children," defined by an administrative age limit, was flawed. The 17/18th birthday is legal/administrative in nature and does not correspond to a physiological transition in the body's maturation. The European Union (EU) adopted a comparable "pediatric" legislation that is in force since 2007.[40,41] The really greedy players in the "pediatric drug development" debate were not pharmaceutical companies, but academic researchers who managed to convince the regulatory authorities, the public, and the lawmakers that "children" needed urgently studies to improve their health. The flawed concept of "pediatric drug development" has allowed generations of "pediatric" researchers to pursue careers based on medically pointless and often harmful studies that were and are often performed on young people who *bodily* are no longer children. And for the younger children, too, the required studies were usually massively exaggerated. It is not the first time in history that those who sell themselves as white knights are those whose greed for funds and careers is worst.

The ramifications of these three outlined factors are complex and cover decades of drug development, medical research, history of public health legislation, and more. While medicine as a clinical discipline is well adapted to treat concrete patients with a well-established diagnosis, medicine is only a small part of the wheels that must be set in motion to deal with a pandemic like COVID-19. Regarding specific patients with a specific diagnosis, medicine has blossomed into a precise science. But the more social issues come into play, the more complex and more complicated things become. Medicine treats the physical patient, and not his legal or administrative status.

This book discusses very broad issues whose discussion is necessary to understand cardinal mistakes that were made and will continue in the fight against the COVID-19 pandemic. The flawed age limits for the clinical trials of the COVID-19 vaccines are a concrete challenge. Multiple inflammatory syndrome (MIS) in children (MIS-C) is another one that will be discussed a bit further below. But the general relationship of modern society to drugs and vaccines, to basic research, and to the institutions that develop new drugs and vaccines is just as important. If today's science wants to win back the public's trust, it must ruthlessly pick up on the flawed concepts that have crept in quietly and unnoticed.

These issues include relevant areas of friction at the interface of the key pillars of healthcare, R&D, and firm but flawed believes and illusions. They include the arrogance and ignorance of leading academic medical experts toward the development of drugs and vaccines; the arrogance of politicians toward the pharmaceutical industry; politicians' continuing belief in the fairy tales of noble knights that fight the profit-greedy industry; the desire of academic researchers to present themselves as selfless, self-sacrificing, and neutral in endless TV interviews, while they advance their research projects, publications, and careers as does everybody else. The convenience of journalists who do not manage to scratch the surface of outright lies or clever misrepresentations.

In 2020, the US Center for Disease Control and Prevention (CDC) issued a warning against a multiple inflammatory syndrome (MIS) in children ("MIS-C"). Such delayed multiple inflammations sometimes occur in COVID-19 after the first acute inflammation has subsided. The CDC warning defined this "pediatric" syndrome in patients up to 20 years.[42] Nobody is bodily a child anymore with 20 years. The US legal age limit to buy and consume alcohol is 21 years. But medicine treats the body, not the legal status. Later in 2020, the CDC published another report that MIS also occurs in adults and named this syndrome "MIS-A".[43] In the meantime, almost thousand medical reports listed in PubMed report about "MIS-C".[44] But the terms "MIS-C" and "MIS-A" are flawed. MIS can occur at any age. It might be a macrophage activation syndrome, a systemic inflammation caused by various virus infections.[45] While the CDC has at least given an age definition to the alleged "MIS-C," UK pediatricians went even further. They describe MIS-C as a "pediatric inflammatory multiple syndromes temporally associated with COVID-19" (PIMS-TS) without specifying an age limit.[46] Why should they? A child is a patient seen by a pediatrician. A criminal is someone arrested by a police officer or someone in jail. Innocent arrested persons do not exist in this circular logic. The UK "PIMS-TC" logic regards current structures and definitions as God-given. Such logic can work for a long time. But not forever. In the meantime, even clinical studies on "MIS-C" have been initiated.[44] Our western medicine has not yet learned to critically question warnings and pronouncements from venerable institutions like the CDC, even if they are obviously nonsensical. Many diseases have different names if they are diagnosed before or after a birthday. In the US, there is a "Center for Adults with Pediatric Rheumatic Illness" (CAPRI).[47] This center's name is logically a contradiction in itself, but it describes one of many bizarre realities of today's life. Some institutions profit from separate "pediatric" studies in rheumatic diseases. For diseases for which FDA-approved effective drugs exist, additional placebo-controlled clinical studies in "children" are demanded by the FDA and even more by the EMA.[40]

In the COVID-19 pandemic, FDA and EMA published jointly a statement about their "pediatric" requirements, defining "children" as anybody up to 17-years-old.[48] Teenagers of 17 years are bodily no longer children. Could the authorities not have used their time for something reasonable? The FDA has given emergency approval for the first drug to easy COVID-19, remdesivir,[49] and has issued reasonable dosing instructions for minors down to babies weighing 3.5 kg.[50] The EMA, however, demands separate "pediatric" remdesivir studies in "children" from birth to 17 years, as if minors were another species.[51]

There are at least two additional elements in the long list of frictions at the interface of healthcare, R&D, drug approval, and clinical medicine. One is the public's trust in authorities. The belief that doctors and scientists selflessly fight for the truth and deeper knowledge is widespread, more so in Europe than in North America, mirrored by the self-picture of scientists, medical doctors, and other healthcare professionals. In contrast to this stands the demonization of the pharmaceutical industry as profit-driven and greedy.[30-38] The industry has almost got used to the constant fire of nonsensical allegations. But this also made the industry defensive.

There was no reasonable reason not to vaccinate 15-year-olds, and to demand separate "pediatric" studies for minors aged 12–15 years.

At present, the authorities only approve drugs and vaccines for those age groups of people who have participated in the clinical studies. For companies, the inclusion of minor patients into pivotal trials means more administrative workload with informed consents that must be signed by the parents. Regulatory authorities followed the advice of "specialists." But meanwhile, it emerges that the specialists' demands for "pediatric" studies are exaggerated and blur the difference between different meanings of the term "child," its bodily vs. its administrative meaning.[40,44]

Who is to blame for the current situation of the non-vaccination of minors? The companies that set the age limits for their studies? The regulatory agencies? The specialists? All of them? None? Which came first, the hen or the egg? Under normal circumstances, years of deliberations would now be scheduled with terribly important professors (TIPs), with a politically correct percentage of female and male professors, an adequate percentage of minorities, etc. The COVID-19 pandemic could allow for once the luxury of thinking about a pragmatic solution. The authorities could lower the age of approval to an extent justified by our understanding of young persons' bodies, e.g., now down to 8 years of age. With 8 years, minors are still pre-pubertal, but their body can already be compared to adults.[52,53] In 94.5% of "pediatric" clinical studies the doses in adolescents aged 12–17 years were identical to those in adults, anyway.[54] Furthermore, the authorities could allow to development of pragmatic dosing recommendations in an "opportunistic" framework for younger persons who are vaccinated and/or treated.[40,44,55] But that will only happen if the organized representatives of medicine, pediatric medicine, pharmacology, development pharmacology and more take a stand and put the authorities under pressure. To do so, they will need to be pushed by the public. A few interested critical journalists would be very helpful. At present, pediatric researchers are demanding "pediatric" COVID-19 studies.[56,57] But the 18th birthday does not change the body. To make the long story short: All pillars of healthcare, public health, and healthcare politics will have to move out of their comfort zone to some degree. One reason for our slowness in fighting the COVID-19 pandemic is convenience, and the unwillingness to address burning issues.

The CDC should correct its warnings against "MIS-C."[42] There is nothing "pediatric" in MIS. The CDC itself has already reported about "MIS-A."[43] MIS is a multiple inflammatory syndrome that rarely occurs in patients of all ages after the first round of a COVID-19 infection.[38,39] "MIS-C" shows another side of today's competitive medical system. Almost thousand scientific papers have been published about "MIS-C."[44] Scientists and medical doctors need publications for their careers. Reports about "MIS-C" attract attention. But they rely on a flawed governmental warning.

Should Western countries order the Russian vaccine Sputnik V, which was approved in Russia based on a phase 1/2 study?[58] Sputnik V is formulated as frozen (storage temperature − 18°C) or freeze-dried (storage temperature 2–8°C).[59] The Russian approach is best described as a strategy of Russian roulette.[60] After an initial phase 1 and 2 study, all sorts of havoc can still occur. Western companies know that unexpected adverse events would ruin the company. The Russian government can afford such an approach. It has powerful police and military at its disposal which, if necessary, stifle protests in blood, poison political opponents, and throw critics in jail.

Should Western countries order a Chinese vaccine? AD5-nCOV (Convidicea), a single-dose vaccine, was developed by the Chinese company CanSino Biologics. CoronoVac, a two-dose vaccine, was developed by another Chinese company, Sinovac. AD5-nCOV (Convidicea) does not need extreme cold conditions for storage and needs only one shot.[61] CoronoVac, which requires two shots, does not need to be frozen.[62] A high-level Chinese official has recently admitted that the Chinese vaccines' efficacy is low with barely above 50%. Steps to "optimize" the vaccination process could include changing the number of doses and the length of time and might include combining different vaccines for the immunization process.[63] See also Chapters 7 and 16.

The free world's media inform us about the tension between Russian and Chinese official declarations, and how their governments trample on human rights where they think they can live with the mild admonitions of the West. We know how deeply the governments of these countries lie. Can we trust the data from their clinical research? Welcome to the real world behind official declarations.

Will we later see confirmed the value of the lengthy FDA and EMA procedures in the examination and approval of drugs and vaccines? Could we further accelerate the development and approval of vaccines and drugs beyond the speed already achieved by OWS?

Western society has technically managed to develop effective vaccines. Compared to former developing periods, this was a true revolution. The competition of the market is brutal and chaotic. OWS has given sufficient funds to several companies to enable the development of effective vaccines in a much shorter time. Governments can create the conditions for companies to develop drugs and vaccines, but the work must be done by such companies themselves. Governments base their decisions on specialists. Do they consult the right ones?

During the first COVID-19 wave, retirement homes isolated residents in a way that resembled solitary confinement. It was worse for the mentally handicapped. In an emergency, there is no time to discuss details. But now we need to address where we have missed common sense and followed the advice of the wrong specialists.[64–67] In the HIV epidemic, the FDA was accused to protect HIV patients to death. We need to ask the same question for COVID-19. The EU is brimming with arrogance, with a medical agency that needs to be kicked in the butt to be moved to the most elementary steps. Is that what was promised in the Lisbon Agenda 20 years ago that aimed to make Europe by 2010 "the most competitive and the most dynamic knowledge-based economy in the world"?[68] It was succeeded by the Europe 2020 strategy with more empty promises.[69]

In the COVID-19 pandemic, state structures and institutions continued with business as usual. The CDC has issued a flawed warning about "MIS-C," which was taken up, repeated, and even translated into pointless clinical studies worldwide.[42,44] Both the FDA and the EMA have continued with their "pediatric" activism,[48] the EMA demands double-blind, placebo-controlled "pediatric" studies in patients up to 17 years, although many of them are bodily no longer children.[40,44,55] The vaccination campaigns are not progressing as fast as it would technically be possible. The COVID-19 pandemic will keep us busy for a while. We will need new vaccines against the new COVID-19 variations and mutations.

Current thinking, traditions, the thinking of institutions, structures, and leaders translate into power that maintains the traditional structures, hindering a resolute approach to the existing challenges. Without public pressure scientists, doctors, authorities, and politicians will not admit mistakes. The previous US president's arrogance and complacency cost him his re-election. The EU is not yet experiencing sufficient pressure to stand by the mistakes it has made in the past and to effectively address the urgent problems of the COVID-19 pandemic. We all have to learn to balance the statements of terrible important professors (TIPs) and top managers of pharmaceutical companies. We should not blindly trust one side or the other. Without pressure, the EU leadership will continue to thrash empty phrases, will continue to announce politically correct-sounding projects, and will make progress too slowly. We live in interesting times.

Next epidemics and/or pandemics will certainly come. They might be more aggressive, might kill more young persons, or have other threatening features. We better open our eyes now. It is always better to be well prepared.

References

1. FDA. *Emergency Approval Pfizer BioNTech COVID-19 Vaccine.* https://www.fda.gov/emergency-preparedness-and-response/coronavirus-disease-2019-covid-19/pfizer-biontech-covid-19-vaccine.
2. FDA. *Emergency Approval Moderna COVID-19 Vaccine.* https://www.fda.gov/emergency-preparedness-and-response/coronavirus-disease-2019-covid-19/moderna-covid-19-vaccine.
3. FDA. *Emergency Approval Janssen COVID-19 Vaccine.* https://www.fda.gov/emergency-preparedness-and-response/coronavirus-disease-2019-covid-19/janssen-covid-19-vaccine.

4. FDA. *Coronavirus (COVID-19) Update: FDA Authorizes Pfizer-BioNTech COVID-19 Vaccine for Emergency Use in Adolescents in Another Important Action in Fight Against Pandemic*; 2021. https://www.fda.gov/news-events/press-announcements/coronavirus-covid-19-update-fda-authorizes-pfizer-biontech-covid-19-vaccine-emergency-use.
5. Stasi C, Fallani S, Voller F, et al. Treatment for COVID-19: an overview. *Eur J Pharmacol.* 2020;889:173644. https://www.ncbi.nlm.nih.gov/pmc/articles/PMC7548059/pdf/main.pdf.
6. Anderson M. *Eight Drugs Trump Has Been Given for his COVID-19 Treatment*; October 5, 2020. https://www.beckershospitalreview.com/pharmacy/8-drugs-trump-has-been-given-for-his-covid-19-treatment.html.
7. US National Instituts of Health (NIH). *National Library of Medicine (PubMed)*. https://pubmed.ncbi.nlm.nih.gov/.
8. Cihlar T, Fordyce M. Current status and prospects of HIV treatment. *Curr Opin Virol.* 2016;18:50–56. https://reader.elsevier.com/reader/sd/pii/S1879625716300207?token=9F194442015CB9D387ADDBEFD1DA893AD1257576FEAE2C93F-07C06F3F46554FC7EA061E5D2DF79C08E0CA6FD3A8F6058.
9. Rappuoli R, Pizza M, Del Giudice G, et al. Vaccines, new opportunities for a new society. *Proc Natl Acad Sci U S A.* 2014;111(34):12288–12293.
10. Wikipedia. *Smallpox Vaccine*. https://en.wikipedia.org/wiki/Smallpox_vaccine.
11. Hilts PJ. *Protecting America's Health: The FDA, Business, and One Hundred Years of Regulation*. New York, USA: Alfred A Knopf Publishers; 2003.
12. Wikipedia. *Operation Wharp Speed*. https://en.wikipedia.org/wiki/Operation_Warp_Speed.
13. King J, Goldenberg D, Goldstein G, et al. *Congressional Budget Responses to the Pandemic: Fund Health care, Not Warfare*; 2021. https://www.ncbi.nlm.nih.gov/pmc/articles/PMC7811077/pdf/AJPH.2020.306048.pdf.
14. Frieden TR, Rajkumar R, Mostashari F. We must fix US health and public health policy. *Am J Public Health.* 2021;111(4):623–627. https://www.ncbi.nlm.nih.gov/pmc/articles/PMC7958072/pdf/AJPH.2020.306125.pdf.
15. *EU Commision on COVID-19 Funding Research*. https://ec.europa.eu/info/live-work-travel-eu/coronavirus-response/overview-commissions-response_en.
16. Wikipedia. *Pfizer–BioNTech COVID-19 Vaccine*. https://en.wikipedia.org/wiki/Pfizer%E2%80%93BioNTech_COVID-19_vaccine.
17. Goldman M. *Viewpoint: Lessons from Operation Warp Speed can help Overcome EU Vaccines Crisis*. Science Business; 2021. https://sciencebusiness.net/covid-19/news/viewpoint-lessons-operation-warp-speed-can-help-overcome-eu-vaccines-crisis.
18. *FDA Approves First COVID-19 Vaccine. Approval Signifies Key Achievement for Public Health*. FDA News Release; 2021. https://www.fda.gov/news-events/press-announcements/fda-approves-first-covid-19-vaccine.
19. *EMA 2021 COVID-19 Vaccines*. https://www.ema.europa.eu/en/human-regulatory/overview/public-health-threats/coronavirus-disease-covid-19/treatments-vaccines/covid-19-vaccines.
20. *Moderna Vaccine Becomes Third COVID-19 Vaccine approved by UK Regulator*. https://www.gov.uk/government/news/moderna-vaccine-becomes-third-covid-19-vaccine-approved-by-uk-regulator.
21. Wikipedia. *COVID-19 Vaccine*. https://en.wikipedia.org/wiki/COVID-19_vaccine.
22. *Covid Vaccine: Why is the EU suing AstraZeneca?* BBC; 2021. https://www.bbc.com/news/56483766.
23. Waterman T. *Why is the European Union suing AstraZeneca again?* CGTN; 2021. https://newseu.cgtn.com/news/2021-05-12/Why-is-the-European-Union-suing-AstraZeneca-again--10bubh0IriU/index.html.
24. Wikipedia. *COVID-19 Pandemic in Israel*. https://en.wikipedia.org/wiki/COVID-19_pandemic_in_Israel.

References

25. Wikipedia. *COVID-19 vaccination in Israel COVID-19 Vaccination in Israel.*
26. *Following Israel's Example: The FDA Recommends the Third Vaccine.* Israel Ministry of Health, Press Release; 2021. https://www.gov.il/en/departments/news/18092021-01.
27. Halpern O. *Israel struggles with COVID surge despite mass vaccinations. Israelis flouting mask requirements may have been a main contributor to the rapid spread of the Delta variant in Israel, experts say.* Aljazeera; 2021. https://www.aljazeera.com/news/2021/8/23/israel-struggles-to-cope-with-surge-of-covid-infections-despite-v.
28. Winer S. *Health Ministry chief says coronavirus spread reaching record heights. As over 10,000 new cases are diagnosed, Nachman Ash tells lawmakers he had hoped recent downward trend would continue.* The Times of Israel; 2021. https://www.timesofisrael.com/health-ministry-chief-says-coronavirus-spread-reaching-record-heights/.
29. DiMasi JA, Grabowski HG, Hansen RW. Innovation in the pharmaceutical industry: new estimates of R&D costs. *J Health Econ.* 2016;47:20–33. https://dukespace.lib.duke.edu/dspace/bitstream/handle/10161/12742/DiMasi-Grabowski-Hansen-RnD-JHE-2016.pdf;sequence=1.
30. Posner G. *Pharma: Greed, Lies, and the Poisoning of America.* New York, NY, USA: Avid Reader Press/Simon & Schuster; 2020.
31. Hirsch ML. Side effects of corporate greed: pharmaceutical companies need a dose of corporate social responsibility need a dose of corporate social responsibility. *Minn J Law Sci Technol.* 2008;9(2). article 7 https://scholarship.law.umn.edu/cgi/viewcontent.cgi?article=1248&context=mjlst.
32. Relman AS. The new medical-industrial complex. *N Engl J Med.* 1980;303(17):963–970. https://www.nejm.org/doi/pdf/10.1056/NEJM198010233031703?articleTools=true.
33. Churchill LR, Perry JE. The "medical-industrial complex". *J Law Med Ethics.* 2014;42(4):408–411.
34. Loder E, Brizzell C, Godlee F. Revisiting the commercial-academic interface in medical journals. *BMJ.* 2015;350, h2957.
35. Steinbrook R, Kassirer JP, Angell M. Justifying conflicts of interest in medical journals: a very bad idea. *BMJ.* 2015;350, h2942.
36. Angell M. *The Truth about the Drug Companies: How they Deceive us and What to Do about it.* New York, USA: Random House Publishers; 2004.
37. Goldacre B. *Bad Pharma: How Drug Companies Mislead Doctors and Harm Patients.* London, UK: Harper Collins Publishers; 2012.
38. Gøtzsche P. *Deadly Medicines and Organised Crime: How Big Pharma Has Corrupted Healthcare.* London, UK: Routledge; 2013.
39. Hirschfeld S. *History of Pediatric Labeling.* https://www.slideserve.com/marlin/history-of-pediatric-labeling.
40. Rose K. *Considering the Patient in Pediatric Drug Development. How Good Intentions turned into Harm.* London: Elsevier; 2020. https://www.elsevier.com/books/considering-the-patient-in-pediatric-drug-development/rose/978-0-12-823888-2. and https://www.sciencedirect.com/book/9780128238882/considering-the-patient-in-pediatric-drug-development.
41. Hirschfeld S, Saint-Raymond A. Pediatric regulatory initiatives. *Handb Exp Pharmacol.* 2011;556:205. 245-268.
42. US Centers for Disease Control and Prevention (CDC) Health Alert Network (HAN) May 14, 2020: HAN00432. *Multisystem Inflammatory Syndrome in Children (MIS-C) Associated with Coronavirus Disease 2019 (COVID-19).* https://emergency.cdc.gov/han/2020/han00432.asp.
43. CDC. *MIS-A Website.* https://www.cdc.gov/mis-c/mis-a.html.
44. Rose K, Grant-Kels JM, Ettienne E, Tanjinatus E, Striano P, Neubauer D. COVID-19 and treatment and immunization of children – the time to redefine pediatric age groups is here. *Rambam Maimonides Med J.* 2021;12(2):e0010.

45. Theoharides TC, Conti P. COVID-19 and multisystem inflammatory syndrome, or is it mast cell activation syndrome? *J Biol Regul Homeost Agents*. 2020;34(5):1633–1636.
46. Harwood R, Allin B, Jones CE, et al. A national consensus management pathway for pediatric inflammatory multisystem syndrome temporally associated with COVID-19 (PIMS-TS): results of a national Delphi process. *Lancet Child Adolesc Health*. 2020. S2352-4642(20)30304-7 https://www.thelancet.com/action/showPdf?pii=S2352-4642%2820%2930304-7.
47. The Brigham and Women's Hospital. *Center for Adults with Pediatric Rheumatic Illness (CAPRI)*. https://www.brighamandwomens.org/medicine/rheumatology-immunology-allergy/services/pediatric-rheumatology-for-adults.
48. FDA/EMA 2020: *FDA/EMA Common Commentary on Submitting an initial Pediatric Study Plan (iPSP) and Pediatric Investigation Plan (PIP) for the Prevention and Treatment of COVID-19*; 2020. https://www.fda.gov/media/138489/download. and https://www.ema.europa.eu/en/documents/other/fda/ema-common-commentary-submitting-initial-pediatric-study-plan-ipsp-pediatric-investigation-plan-pip_en.pdf.
49. FDA. *Remdesivir Prescribing Information*; 2020. https://www.accessdata.fda.gov/drugsatfda_docs/label/2020/214787Orig1s000lbl.pdf.
50. FDA. *Fact Sheet For Healthcare Providers. Emergency Use Authorization (EUA) Of Veklury® (Remdesivir) For Hospitalized Pediatric Patients Weighing 3.5 Kg To Less Than 40 Kg or hospitalized Pediatric Patients Less Than 12 Years Of Age Weighing At Least 3.5 Kg*; 2020. https://www.fda.gov/media/137566/download.
51. EMA. *Remdesivir PIP EMEA-002826-PIP01-20*; 2020. https://www.ema.europa.eu/en/documents/pip-decision/p/0046/2020-ema-decision-29-january-2020-agreement-pediatric-investigation-plan-granting-deferral-granting_en.pdf.
52. Kearns GL, Abdel-Rahman SM, Alander SW, Blowey DL, Leeder JS, Kauffman RE. Developmental pharmacology – drug disposition, action, and therapy in infants and children. *N Engl J Med*. 2003;349(12):1157–1167. https://pdfs.semanticscholar.org/55f8/1745303e9aaec7f4cb85f0b5921eec14a9c0.pdf.
53. Beunen GP, Rogol AD, Malina RM. Indicators of biological maturation and secular changes in biological maturation. *Food Nutr Bull*. 2006;27(4 Suppl Growth Standard):S244–S256. https://journals.sagepub.com/doi/pdf/10.1177/15648265060274S508.
54. Momper JD, Mulugeta Y, Green DJ, et al. Adolescent dosing and labeling since the Food and Drug Administration amendments act of 2007. *JAMA Pediatr*. 2013;167(10):926–932. https://jamanetwork.com/journals/jamapediatrics/fullarticle/10.1001/jamapediatrics.2013.465.
55. Rose K, Neubauer D, Grant-Kels JM. Rational use of medicine in children – the conflict of interests story. A review. *Rambam Maimonides Med J*. 2019;10(3), e0018. Review https://doi.org/10.5041/RMMJ.10371. https://www.rmmj.org.il/userimages/928/2/PublishFiles/953Article.pdf.
56. Kamidani S, Rostad CA, Anderson EJ. COVID-19 vaccine development: a pediatric perspective. *Curr Opin Pediatr*. 2021;33(1):144–151. https://journals.lww.com/co-pediatrics/Fulltext/2021/02000/COVID_19_vaccine_development__a_pediatric.20.aspx.
57. Kao CM, Orenstein WA, Anderson EJ. The importance of advancing SARS-CoV-2 vaccines in children. *Clin Infect Dis*. 2020;72. ciaa712 (Jun 3) https://www.ncbi.nlm.nih.gov/pmc/articles/PMC7314192/pdf/ciaa712.pdf.
58. Logunov DY, Dolzhikova IV, Shcheblyakov DV, et al. Safety and efficacy of an rAd26 and rAd5 vector-based heterologous prime-boost COVID-19 vaccine: an interim analysis of a randomised controlled phase 3 trial in Russia. *Lancet*. 2021;397(10275):671–681. https://www.ncbi.nlm.nih.gov/pmc/articles/PMC7852454/pdf/main.pdf.
59. Wikipedia. *Sputnik V COVID-19 Vaccine*. https://en.wikipedia.org/wiki/Sputnik_V_COVID-19_vaccine.
60. Wikipedia. *Russian Roulette*. https://en.wikipedia.org/wiki/Russian_roulette.

61. Wikipedia. *Ad5-nCoV (Convidicea)*. https://en.wikipedia.org/wiki/Ad5-nCoV.
62. Wikipedia. *CoronaVac* https://en.wikipedia.org/wiki/CoronaVac.
63. BBC News. *Chinese Official says Local Vaccines 'Don't have High Protection Rates'*; 2021. https://www.bbc.com/news/world-asia-china-56713663.
64. Andrist E, Clarke RG, Harding M. Paved with good intentions: hospital visitation restrictions in the age of coronavirus disease 2019. *Pediatr Crit Care Med*. 2020;21(10):e924–e926. https://www.ncbi.nlm.nih.gov/pmc/articles/PMC7314338/pdf/pcc-21-e924.pdf.
65. Anderson EJ, Campbell JD, Creech CB, et al. Warp speed for COVID-19 vaccines: why are children stuck in neutral? *Clin Infect Dis*. 2020;, ciaa1425. https://www.ncbi.nlm.nih.gov/pmc/articles/PMC7543330/pdf/ciaa1425.pdf.
66. Kales HC, Maust DT. Good intentions, but what about unintended consequences? *Drug Saf*. 2017;40(8):647–649.
67. Byrne JM. Administrative ethics: good intentions, bad decisions. *Healthc Manage Forum*. 2018;31(6):265–268.
68. Wikipedia. *Lisbon Strategy*. https://en.wikipedia.org/wiki/Lisbon_Strategy.
69. Wikipedia. *Europe 2020 Strategy*. https://en.wikipedia.org/wiki/Europe_2020.

CHAPTER 2

We are no longer hunters and gatherers. Societies, states, values, and healthcare today

The livelihood of humans for much of early history was hunting and gathering in a nomadic life where most or all food was obtained by collecting wild plants, gathering immobile aquatic animals such as clams, and hunting wild animals. This went on for several hundred thousand years. Things changed rather slowly these days, but change they did. Tools and weapons improved.[1] During the late glacial era, roughly 30,000 years ago, wolves were the first animals that were domesticated by hunter-gatherers, both in Europe and Asia. Gradually, some of the early tamed wolves became dogs.[2,3] Another ten or fifteen thousand years later, the systematic domestication of plants and animals began. Agricultural societies emerged, with societal differentiation, hierarchies, increasingly sophisticated tools, and weapons, initially made of wood and stone, then gradually of metal.

To look at societies based on hunting and gathering is relevant in the context of the COVID-19 pandemic for several reasons. The population density in those times was massively lower than today. There were no tamed horses or other animals to ride on. There were no wagons on which one could transport larger household items, sick persons, or fat and lazy chiefs ☺. Getting around on foot had several massive advantages from the point of view of today's medical challenges. Hunter-gatherer populations were in general in excellent health and are characterized in today's medical literature as models in public health. If you are constantly moving and have no permanent home, you can neither afford to become obese and immobile nor do you have the resources to do so.[4] Periods of abundance alternated with periods of hunger. People whose metabolism could make the most of the food available had better chances of survival. Even in these early times, viral diseases were certainly transmitted from animals to humans. But the framework for potential dissemination was fundamentally different from today. Today's two types of the human immunodeficiency virus (HIV-1 and HIV-2) originated from hunting chimpanzees and gorillas and consuming them as "bushmeat" in Central and West Africa.[5] In apes, the virus is called the simian immunodeficiency virus (SIV). But that is only one among several further parts of the puzzle. Occasionally, people who hunt monkeys or have to do with hunted monkeys become infected with their viruses while slaughtering them, touching raw meat, consuming undercooked meat, or having an open wound on their hands or in the mouth. Several HIV-seronegative Cameroonians were seropositive for SIV, with the seropositivity correlated to the exposure risk.[6] However, this alone would not explain the HIV epidemic. HIV probably emerged epidemically as a result of harsh conditions, forced labor, displacement, and unsafe injection and vaccination practices associated with colonialism in Africa. The workers in plantations, construction projects, and other colonial enterprises were supplied with bushmeat, which contributed to an increase in hunting and a higher incidence of human exposure to SIV. The colonial authorities also vaccinated against smallpox and other diseases. Many injections were performed without the equipment sufficiently sterilized between uses. A further factor may have been the fact that the camp residents had money, the breaking up of traditional living in small hamlets, and the prostitution associated with the forced labor camps. Also, the extreme stress of forced labor could depress the immune system of workers, prolonging the primary acute infection period of persons newly infected with SIV. But all that occurred still rather regionally. HIV spread to the Western Hemisphere probably from central Africa to Haiti, from there to New York and California, and became eventually a global epidemic as we know it today. HIV became a global epidemic that had as a basis the availability of modern forms of transportation, modern mobility, and societies with less strict codes regulating sexuality. The diagnosis of HIV required modern communication. The development of treatments for HIV also required a fundamental shift in the fundamental views of the FDA.[7]

The spread of an epidemic like COVID-19 not only requires a much higher population density and modern means of transport. Depending on the path of the infection, sexual habits come into play, or other social customs such as larger crowds at sporting, religious, and other events, and many other factors. But we are not done yet with the hunters and gatherers. We would not understand modern human history without the history of the domestication of animals and the transmission of diseases from animals to humans. Pathogens shared with wild or domestic animals cause more than 60% of infectious diseases in man.[8-11]

Infections resulting from bites of animals infected with rabies are described in the earliest written evidence.[8-11] Such infections will have occurred also much earlier. Most other infections beyond infecting individuals, however, emerged with the systematic coexistence of humans and animals in agriculture from fifteen to ten thousand years ago on, when closed larger settlements emerged, the herds of animals became larger, and more and more people lived together in relatively small areas. Agriculture emerged when men used berries and seeds to sow and produce more instead of eating them right away. Some plants were selected on the basis of characteristics that in the wild prevented their propagation, such as peas that did not distribute their seeds by small explosions when they had become ripe, or almond trees that were selected because the almonds were not bitter, and thus would have been eaten by animals unless men collected and sowed them. A relatively late step forward was the grafting of trees on wild stems, such as apples, pear, or other fruits. The four major animal and domestic species, cattle, goat, sheep, and pig, emerged in the Middle East and Asia. Poultry was domesticated both in China and India probably 6000 years ago, and turkey in North America 3000 years ago.[2,10,11]

In classical thoughts, life was often thought to have been better in the past. Origin stories from the lost Garden of Eden in the bible to the Golden Age of ancient Greece describe a utopian past where humans lived in harmony with nature and were healthy and well-nourished.[4] The prevailing view of early history assumed that man lived in a social environment of permanent harmony and well-being, based on the writings of the Swiss philosopher, writer, and composer Rousseau, who established the concept of the "noble savage" in the 18th century. His basic assumption was that the first human groups lived in their natural state, in fraternity, well-being, and without conflicts.[12] Also Marx and Engels' nostalgic descriptions of early farming communities all paint pictures of healthful idyll before the corruption wrought by progress and industrialization.[13]

Today, we have become more realistic in looking at early history. There were and are societies that are more prone to war than others. The Chatham Islands, today part of New Zealand, are an archipelago in the Pacific Ocean about 800 km east of New Zealand. Maori settlers from today's New Zealand populated them around AD 1500, becoming gradually the Moriori. The Chathams are colder and less hospitable than the land the settlers had left behind and were unsuitable for the cultivation of most crops known to Polynesians. The Moriori adopted a hunter-gatherer lifestyle. Food was almost entirely marine-sourced. Furthermore, they developed a pacifist culture that avoided warfare and replaced it with dispute resolution by ritual fighting and conciliation. But in 1835, almost a thousand Maori came in a ship of European origin, armed with guns, clubs, and axes, and equipped with a supply of potatoes that they planted there. They killed most Moriori and enslaved those that survived, which were few. Only at the end of the 1860s became the remaining Moriori free from slavery. Today, their culture is enjoying a renaissance, both in the Chatham

Islands and among their descendants on New Zealand's mainland, but their language is no longer spoken, sharing the fate of many other languages.[10,14] Hunters and gatherers have not been always as peaceful as they were romantically hypothesized by Rousseau, nor did they always have a culture of continuous warfare. But in general, and as documented by today's learnings of archeology, warfare and violence have always been part of human life.[15,16]

The expansion of man over the world was in the roughly first 10,000 years not driven by modern states. Instead, societies had different levels of the organization. Written testimonies tell us about life and death in early societies as far as they had already scripture. Already in those days wounds were treated, broken limbs splinted, and alcohol was made by fermentation in waterproof vessels.[17]

Yeast converts sugar into ethanol and CO_2 via fermentation. Yeasts have been used for thousands of years by mankind for fermenting food and beverages. In the Neolithic times, fermentations were certainly initiated by naturally occurring yeasts. Eventually, humans started to consciously add selected yeast to make beer, wine, or bread, giving, inter alia, rise to the creation of new species of yeast. Although the evolutionary history of wine yeast has been well described, the histories of other domesticated yeasts still offer plenty of space for further research.[18]

Agriculture advanced the worldwide spread of human populations and led to the invasion of areas by cultivated plants and livestock animals. As a result, fewer animal and plant species are today found throughout the world than thirty, twenty and ten thousand years ago. In the absence of predators and parasites, invaders often undergo explosive population increases, with severe consequences for the crop plants and domesticated animals. A fungal infection causing potato blight was introduced into Europe from America in the 19th century when potatoes, imported from America to Europe a few centuries earlier, had become an essential part of the diet of many European countries. The causative fungus probably came from America as well. It invaded large cultivated areas of Europe and was the cause of the Great Famine in Ireland. Another example is the insect phylloxera, originally native to North America, which devastated most vineyards in Europe and worldwide in the 19th century.[19] The only successful means of controlling phylloxera was the grafting of phylloxera-resistant American rootstock to European types of vine.[20]

Man is a social being. Although humanity is made up of many individual people, there is much that goes beyond the individual. There is the spoken language, there are the different conventions of clothing, there are permitted, less permitted, and forbidden types of body language, and much more. Religion can be described as a social and cultural system of behaviors, practices, morals, worldviews, sanctified places, prophecies, and ethics that relate humanity to supernatural, transcendental, and spiritual elements. There is no scholarly consensus on how to precisely define religion.[21,22] Religion must have been present in early humans. Archeologists take apparent intentional burials of early man and Neanderthals from as early as 300,000 years ago as evidence of religious ideas. Other evidence, mostly later, of early religious ideas are artifacts from the Stone Age, including figurines of hybrids of man and animals, figurines of women, cave paintings, and more.[22] As usual, the scientific interpretation if and how they relate(d) to religious ideas is controversial. Religious rituals are both verbal and non-verbal, including dance, music, rituals, stating sacred truths, elders or shamans going into a trance, with or without the help of intoxicating substances, and more.[22,23] If religion and other systematic thought buildings came before or after the emergence of language is hotly debated,[22–24] a debate that is reminiscent of the discussion about which came first, the hen or the egg. Can we prove that early men had sex? Not by photography or written documents. But common sense tells us that we would not have survived without it.

Analyzing both the religious beliefs of still existing hunter-gatherers, and looking into the history of religious thoughts allow some understanding of the beliefs that hunter-gatherers and early communities had. They saw the world in an animistic way, with objects, places, animals, plants, rocks, rivers, weather systems, and more possessing distinct spiritual essences. There was a belief in an afterlife. They had persons with the special task for communication with supernatural, transcendental, and spiritual elements; they worshipped their ancestors and more.[22–24] We still have remnants of animism in modern English: Sunday is the day of the sun and its god, Monday the day of the moon and its god, Thursday the day of the Norse god of Thunder, Thor.

Today's predominant monotheistic religions emerged as thought structures with the hierarchically structured, complex forms of society. It is essential to visualize the minds of hunters and gatherers and their societies to understand how they saw disease. Fractures, burns, cancer, animal bites, difficult childbirths, and more have accompanied mankind since its earliest origins. There were no trained doctors or pharmacists. There were persons with experience in how to deal with wounds, infections, childbirth, and more. The world was perceived as something where spiritual and material elements were not as clearly separated from each other as we perceive them today. Healing was partially perceived as the job of those who specialized in dealing with the gods. Medicine and religion were not clearly divided. Learnings were achieved by trial and error. These learnings were passed down through generations if the respective tribe or group survived. Otherwise, the accumulated knowledge got simply lost. Fruits, leaves, bark, roots, and animal parts were rubbed onto the body, boiled, drunk, or eaten, and their effects were noted and discussed. Tribes met and engaged in warfare, trade, or alternately both. Long before recorded time, salt trade routes were established.[17] Other trade routes emerged with the transition from the stone age to early forms of metal and finally with the development of empires in Europe and Asia that used metals both for weapons and for tools of peaceful daily life.

Societies of hunters and gatherers did not live in isolation. They were not as highly mobile as we are today. The groups also lived in comparable sanitary conditions. Pathogens that infected one group sooner or later also infected neighboring groups – unless a completely new group entered the equation. When Cortés and his small army conquered Mexico, he, his priests, and his soldiers brought with them the pathogens that had taken root in man in Europe and Asia over the past millennia. The immune system of the native Americans was unprepared for these pathogens that killed many and further weakened their military defense capabilities.

Some classical endemic diseases can today be detected by archeology that uses genetic testing. Smallpox affected people in ancient Egypt, including pharaohs. Other endemic diseases included the pest, carried by lice on rats, and diseases such as typhoid or cholera that often affected armies in their operations.

All classical endemic diseases that we find described in ancient writings affected already rather complex societies that had written documentation. In a few cases, such as the European invasions in the Americas, the invaders themselves had people with them who could read and write and document what they observed. Those days, there were no hospitals and no cures. The doctors, priests, and the scribes were helpless against infectious diseases which they did not see as infectious diseases but as signs or punishments from one god or several gods, depending on the god or gods they believed in.

Although the average life expectancy calculated by today's standards was low, especially due to the high child mortality rate, there were people in prehistory who lived to become rather old. We now assume that most people who had once turned 45 lived on for decades.[4] But first of all, there was no one and no institution at the time that would have calculated the average life expectancy. The term "life expectancy" is a term used in today's highly complex society. It required public administration, statistics, modern communication, and more. Secondly, a disease like COVID-19 that is survived by most and kills predominantly old and very old people would probably not even have been acknowledged as a disease. In contemplating history, we must be aware that we often impose our current categories on circumstances that were completely different then. That should not prevent us from making such considerations. But we have to put our conclusions into perspective.

When interest in earlier societies grew in the past centuries, the prohibition of incest, i.e. sexual intercourse with close blood relatives was perceived by social scientists and historians as a universal trait found in every known culture.[25] Incest is indeed one of the most widespread of all cultural taboos.[26] Because incest so closely intertwined with taboo until the recent past close evaluation was thwarted by a nearly implacable defensive shield within families who wanted to maintain their social reputation. Adding to this problem were that clinicians and others considered incest as being nontraumatic and no cause for concern.[27,28] Today, it is acknowledged as something that massively harms young people. Young persons who have been sexually assaulted by family members based on their position of physical and social power often suffer from it all their lives. Thus, we place today increasingly emphasis on preventing child abuse. But a critical look at history shows that the prohibition of incest was not as absolute and universal as was assumed for a while. In Greek and other classical myths of antiquity, the gods commit incest abundantly.[25–28] In the upper strata of the meanwhile more differentiated societies, incest was often used to keep the bloodlines of leading families pure, including those of the Pharaohs, the Inca rulers, and more. Today's genetic analyzes allow an insight beyond the traditional stories, myths, and written documents. Researchers identified the adult son of a first-degree incestuous union from remains near an elaborate neolithic monument in Ireland. The authors emphasize that such socially sanctioned matings are documented almost exclusively among politico-religious elites, specifically within polygynous and patrilineal royal families that were headed by god-kings.[29] Like so many other assessments of early science, a closer look at the prohibition of incest shows that a large part of the early assessments was based less on the data collected and more on contemporary ideas and prejudices that were projected and interpreted into the observed societies. And here is the parallel to looking at the COVID-19 pandemic versus previous epidemics. Many of the comparisons made today are based too much on the assumption that basic elements of contemporary life are taken for granted.[30]

Probably the most important difference between early societies and today is the high value we give to every individual life. This high value has become an elementary part of the basis of today's human coexistence. When in early history someone got sick, it was his personal fate and the will of the gods whether he survived or not. Nowadays social mechanisms are switched on. An ambulance or a doctor, or both are called. If that is not enough, the sick person is taken to the hospital and treated there. The COVID-19 pandemic is far less lethal for individual patients compared to other classic pandemic diseases,[30] but it is also fundamentally different. Politicians who would openly declare that they do not care how many people die would have to fear for their re-election. The highest thing a corrupt or otherwise meschugge politician can do today is to belittle the pandemic. An alternative is open lies up to falsifying official statistics. Tanzania is one of the countries that has continually claimed that COVID-19 does not exist in them. The president of Tanzania has officially died of a heart attack. But it is an open secret that he died of COVID-19.[31] Hypocrisy is one of the many basic constants of our modern society today, to which the many means of modern mass communication are available. Some lies and false ideas last a while longer in today's large states and the global community than in a small, limited social environment. But sooner or later the old folk wisdom still come true, including that lies have short legs.

References

1. Wikipedia. *Hunter-Gatherer.* https://en.wikipedia.org/wiki/Hunter-gatherer.
2. Vigne JD. The origins of animal domestication and husbandry: a major change in the history of humanity and the biosphere. *C R Biol.* 2011;334(3):171–181. https://www.sciencedirect.com/science/article/pii/S1631069110002982?via%3Dihub.
3. Tresset A, Vigne JD. Last hunter-gatherers and first farmers of Europe. *C R Biol.* 2011;334(3):182–189. https://www.sciencedirect.com/science/article/pii/S1631069110002994?via%3Dihub.
4. Pontzer H, Wood BM, Raichlen DA. Hunter-gatherers as models in public health. *Obes Rev.* 2018;19(Suppl 1):24–35. https://onlinelibrary.wiley.com/doi/epdf/10.1111/obr.12785.
5. Wikipedia. *History of HIV and AIDS.* https://en.wikipedia.org/wiki/History_of_HIV/AIDS.
6. Kalish ML, Wolfe ND, Ndongmo CB, et al. Central African hunters exposed to simian immunodeficiency virus. *Emerg Infect Dis.* 2005;11(12):1928–1930. https://www.ncbi.nlm.nih.gov/pmc/articles/PMC3367631/pdf/05-0394.pdf.
7. Hilts PJ. *Protecting America'S Health: The FDA, Business, and One Hundred Years of Regulation.* New York, USA: Alfred A Knopf Publishers; 2003.
8. Karesh WB, Dobson A, Lloyd-Smith JO, et al. Ecology of zoonoses: natural and unnatural histories. *Lancet.* 2012;380(9857):1936–1945. https://www.ncbi.nlm.nih.gov/pmc/articles/PMC7138068/pdf/main.pdf.
9. Åsjö B, Kruse H. Zoonoses in the emergence of human viral diseases. *Perspect Med Virol.* 2006;16:15–41. https://www.ncbi.nlm.nih.gov/pmc/articles/PMC7114646/pdf/main.pdf.
10. Diamond J. *Guns, Germs, and Steel: The Fates of Human Societies.* New York, NY, USA: Vintage Books; 1998.
11. Wikipedia. *Zoonosis.* https://en.wikipedia.org/wiki/Zoonosis.
12. Wikipedia. *Jean-Jacques Rousseau.* https://en.wikipedia.org/wiki/Jean-Jacques_Rousseau.
13. Wikipedia. *Primitive communism.* https://en.wikipedia.org/wiki/Primitive_communism.
14. Wikipedia. *Moriori.* https://en.wikipedia.org/wiki/Moriori.
15. Keeley LH. *War before Civilization: The Myth of the Peaceful Savage.* London, UK: Oxford University Press; 1996.
16. Walker PL. A bioarchaeological perspective on the history of violence. *Annu Rev Anthropol.* 2001;30:573–596. https://unlcms.unl.edu/anthropology/hames-home-page/courses/war/walker-bioarch-war.pdf.
17. Hacker M. History of pharmacology - from antiquity to the twentieth century. In: Hacker M, Messer WS, Bachmann KA, eds. *Pharmacology: Principles and Practice.* London, UK: Academic Press (Elsevier); 2009 [Chapter 1] https://www.amazon.com/Pharmacology-Principles-Practice-Miles-Hacker-ebook/dp/B004Y6ZBXW#reader_B004Y6ZBXW.
18. Sicard D, Legras JL. Bread, beer and wine: yeast domestication in the Saccharomyces sensu stricto complex. *C R Biol.* 2011;334(3):229–236. https://www.sciencedirect.com/science/article/pii/S1631069110003057?via%3Dihub.

19. Guillemaud T, Ciosi M, Lombaert E, et al. Biological invasions in agricultural settings: insights from evolutionary biology and population genetics. *C R Biol.* 2011;334(3):237–246. https://www.sciencedirect.com/science/article/pii/S1631069110002970?via%3Dihub.
20. Wikipedia. *Phyloxxera.* https://en.wikipedia.org/wiki/Phylloxera.
21. Peoples HC, Duda P, Marlowe FW. Hunter-gatherers and the origins of religion. *Hum Nat.* 2016;27(3):261–282. https://www.ncbi.nlm.nih.gov/pmc/articles/PMC4958132/pdf/12110_2016_Article_9260.pdf.
22. Wikipedia. *Religion.* https://en.wikipedia.org/wiki/Religion.
23. Wikipedia. *History of Religion.* https://en.wikipedia.org/wiki/History_of_religion.
24. Wikipedia. *Evolutionary Origin of Religions.* https://en.wikipedia.org/wiki/Evolutionary_origin_of_religions.
25. Martin EJ. Incest/child sexual abuse: historical perspectives. *J Holist Nurs.* 1995;13(1):7–18.
26. Wikipedia. *Incest.* https://en.wikipedia.org/wiki/Incest.
27. Lande RG. Incest. Its causes and repercussions. *Postgrad Med.* 1989;85(8):81–86. 89, 92.
28. Serrano AC, Gunzburger DW. An historical perspective of incest. *Int J Fam Ther.* 1983;5(2):70–80. https://link.springer.com/content/pdf/10.1007/BF00924434.pdf.
29. Cassidy LM, Maoldúin RÓ, Kador T, et al. A dynastic elite in monumental Neolithic society. *Nature.* 2020;582(7812):384–388. https://www.ncbi.nlm.nih.gov/pmc/articles/PMC7116870/pdf/EMS114623.pdf.
30. Pitlik SD. COVID-19 compared to other pandemic diseases. *Rambam Maimonides Med J.* 2020;11(3), e0027. https://www.ncbi.nlm.nih.gov/pmc/articles/PMC7426550/pdf/rmmj-11-3-e0027.pdf.
31. Odula T. *Tanzania gives Hero's Burial to President who Denied Virus.* Associated Press (AP) News; March 26, 2021. https://apnews.com/article/dar-es-salaam-tanzania-coronavirus-pandemic-africa-john-magufuli-6f5b05e16e218c77e1b4734a194be55a.

CHAPTER 3

Development of drugs and vaccines

The disease is as old as humanity. We see the traces of illness in ancient fossil remains, skeletons in museums, and other holdovers from prehistory. Injured and sick people could count on the help of their tribe, although most details are lost in the fog of prehistory. Health challenges included injuries, malnutrition, infectious diseases, malignancies, and many more. In prehistory, the profession of healing and religious care was not yet strictly separated. All forms of ancient medicine shared a supernatural orientation and a belief in magic. Disease and misfortune were attributed to supernatural agents. Diseases for which there was no obvious immediate causes that were assumed to be caused by ghosts, spirits, gods, sorcery, witchcraft, or the loss of one of the victim's special souls.[1] Those involved with maintaining health knew the values of different foods, medicinal plants, minerals, and the healing properties of smoke and chants.[2]

Until the 18th century, the use of herbal and other medicines had been entirely based on empiricism. Only in the late 18th century the foundations of pharmacology, the study of the actions of drugs and how they exert their effects, begin to emerge.[3] In 1900, life expectancy in the US was 47 years. Infectious diseases such as pneumonia, influenza, tuberculosis, diphtheria, smallpox, pertussis, measles, and typhoid fever were the leading causes of mortality, with children and young adults most affected. Vaccination, together with improved hygiene, improved housing, nutrition, and effective drugs played a key role in the 20th century in drastically reducing the mortality from infectious diseases. Vaccination can be regarded as the most effective medical intervention ever introduced. Together with clean water and sanitation, it has eliminated a large part of the infectious diseases that once killed millions of people.[4] Vaccines are biological preparations that help the body develop immunity to a particular infectious disease. Traditionally, vaccines were developed by attenuating or inactivating the respective pathogen. Such vaccines led to the eradication of smallpox and significantly restricted diseases such as polio, tetanus, diphtheria, and measles.[5] Today's vaccine development has become an industrial business together with the development of new drugs. It uses all methods of modern life science to induce immunity to infectious diseases, including vaccines based on recombinant DNA, viral-like particles identical to those released by the respective virus, and the conjugation of bacterial capsular polysaccharides to carrier proteins to make polysaccharide antigens T-cell dependent and therefore immunogenic in infants. Furthermore, the traditional boundaries between vaccines and drugs are merging. Infusion with autologous T cells engineered in vitro to express high-affinity chimeric antigen receptors (CARs) targeting the CD19 antigen present in B cells can proliferate and efficiently eliminate aggressive, treatment-refractory leukemia cells from patients with chronic lymphocytic leukemia (CLL) and acute lymphoblastic leukemia (ALL), resulting in complete tumor remission. Research is underway on vaccines for the prevention of and therapy of cancer and neurodegenerative diseases.[4,6]

Modern drug development has several historical and structural roots. One is industrial research & development. Today's pharmaceutical industry traces its origin to two major types of companies: those that had previously supplied natural products and moved into large-scale production of drugs in the middle of the 19th century. The other ones were newly dye and chemical companies that established research laboratories and discovered medical use for their products.[3] A further key element in modern science, which is continuously changing and advancing the foundations for research into new active ingredients and priniciples. It began with the academic discovery of penicillin, continued with the immune system's involvement in the life-saving treatment of ALL,[7,8] and led recently to the novel principle of vaccines against COVID-19 based on the use of messenger RNA (mRNA), which required several groundbreaking medical and scientific discoveries.[9,10] Two final crucial elements are the regulatory agencies, without whose consent drugs cannot be marketed today, and the prescription of approved drugs by medical doctors. The prescription of drugs by medical doctors only emerged in the wake of the scientific and therapeutic breakthroughs during and after World War II.[11–13]

Clinical studies for the approval of drugs and vaccines have to a large degree replaced individual observations by medical doctors and pharmacists by systematic collective data collection and its documentation. For this system to work, trust in science and institutions is essential.

The development of drugs and vaccines is today the process new medicines must go through until they are approved by the respective regulatory authority, can be prescribed, sold, and used. New medicines can be new molecules or, increasingly today, biologically defined substances such as monoclonal antibodies, receptor antagonists, or substances that interrupt or slow down unwanted passways in the body. Early vaccines were developed by attenuating or inactivating the respective pathogen. Today, the development of vaccines and drugs has become a high-tech business applying the latest progress in understanding genetics and the immune system.

In 1977, the ulcer medication Tagamet became the first-ever pharmaceutical blockbuster, earning its manufacturer more than US$ 1 billion a year. It was followed by a succession of blockbuster products, each financially more successful than its predecessors. Prozac, the first selective serotonin reuptake inhibitor (SSRI) antidepressant was launched in 1987. Omeprazole, the first proton pump inhibitor (PPI), was introduced in 1989. Atorvastatin, marketed as Lipitor in 1996, became the world's best-selling drug of all time, exceeding US$ 125 billion in sales over 15 years approximately.[3]

The industrial development of penicillin advanced during World War II with the support of the US Office of Scientific Research and Development (OSRD), an institution that today is mostly mentioned in connection with the "Manhattan Project" that developed the first nuclear bombs. Penicillin did not begin to play a major role in medicine due to publications in peer-reviewed journals. Its breakthrough role began with its industrial production.[11]

The introduction of thalidomide and the realization that it was responsible for early death and deformities in many thousands of children worldwide,[14] led to a fast and determined US reaction. In 1962, a new law authorized the FDA to only approve drugs whose efficacy and safety had been proven. It took decades for the rest of the world to follow this precedence, but today it is in force worldwide.[11,15]

The technical steps in the development of drugs and vaccines are well described by mainstream science and on the websites of the major regulatory agencies. In the US this is the Food and Drug Administration (FDA), in the EU the European Medicines Agency (EMA), and in Japan the Pharmaceuticals and Medical Devices Agency (PMDA).[16,17] This development process is divided into preclinical (in vitro and in vivo testing in animals) and clinical (clinical trials in human subjects) stages.[18] The FDA lists five key steps: (1) Discovery and Development; (2) Preclinical Research; (3) Clinical Research; (4) FDA Review; (5) FDA-Post-Market Safety Monitoring.[19] The EMA lists three: (1) research & development; (2) marketing authorization; (3) post-authorization.[20]

Most phase 1 studies in drugs and vaccines are performed in healthy volunteers.[18]

Clinical studies have to a considerable degree replaced individual observation of the effect of medicines by clinical practitioners with collective, organized observation. Most studies that aim at approval by a regulatory authority are performed in many centers (multicenter studies). Once enough patients have been treated, the study leadership examines which treatment worked better. As long as for a specific drug it is not known it if works at all, and if the observed results are based on true efficacy or only on wishful thinking, examination against placebo is usual. Placebo (Latin for "I will please") is a pill, a tablet, or an injection that from its external appearance cannot be differentiated from the drug under examination. "Randomized" treatment means that neither the patient nor the treating doctor knows which medication the respective patient receives. Would a doctor know that blue pills contain the true medication and white pills just placebo, he might be tempted to treat patients he perceives as attractive with blue pills, and other ones with white ones. He would not even have to be aware of this.

Unlike drugs, which are given to patients, vaccines are received by healthy individuals. While pivotal drug studies usually recruit hundreds to thousands of patients, the number is much higher in vaccination studies, where already phase II studies recruit hundreds to thousands of patients, and phase III trials often thousands to tens of thousands of subjects.[18] The interim safety and efficacy analysis of the AstraZeneca US COVID-19 phase III trial of AZD1222 was based on 32,449 participants with a 2:1 randomization of vaccine to placebo.[21] In the BioNTech BNT162b2 trial 43,448 subjects received either BNT162b2 or placebo in a 1:1 randomization.[22]

A competitive environment is one of the reasons that today more and more drugs are developed against rare diseases. For the frequent "classical" diseases such as hypertension, dyslipidemia, gastric ulcers, or esophagitis there are today already many effective generics on the market. A new development for yet another antihypertensive drug would no longer be financially rewarding. But there are also further reasons for the development of drugs for rare diseases, specifically US and EU laws that incentivize the development of new drugs against rare diseases. The COVID-19 pandemic is yet another special case and made the development of new effective vaccines more urgent than ever.

Today's standard in drug and vaccine development is the result of hard disputes, and not just between drug manufacturers and authorizing authorities. It was an argument in which not only science was involved, but also the entire public through the reports and discussions in the free press. In the early years of drug development, drug manufacturers did everything possible to suppress the spread of knowledge about side effects as they emerged. The members of the sales force were instructed to lie openly in response to inquiries from doctors about side effects that the company in question had long known. Over time, the public outrage generated by reports of such machinations was reflected in the US in tougher laws and increased FDA oversight of drug development. The development of such stricter surveillance was, of course, facilitated by the general hostility towards the pharmaceutical industry. Worldwide, the US with its FDA has developed a clear leading position in drug development, which was then followed by agencies such as the EMA and the PMDA.[11,15]

How difficult it is to develop a culture of precise and comprehensive documentation of drug development and its strict monitoring by the regulatory authorities was exemplified during the COVID pandemic. The clinical data on the development of the Russian and Chinese vaccines have never been disclosed comparably to the scientific documentation and its publication of the vaccines developed by AstraZeneca, Pfizer-BioNTech, Moderna, and Janssen. In consequence, the trust in these vaccines in the professional world is rightly limited; see also Chapters 7 and 17.

References

1. Magner LN, Kim OJ. *A History of Medicine*. 3rd ed. Boca Raton, FL, USA: CRC Press, Taylor & Francis; 2018.
2. Hacker M. History of pharmacology - from antiquity to the twentieth century. In: Hacker M, Messer WS, Bachmann KA, eds. *Pharmacology: Principles and Practice*. London, UK: Academic Press (Elsevier); 2009 [Chapter 1] https://www.amazon.com/Pharmacology-Principles-Practice-Miles-Hacker-ebook/dp/B004Y6ZBXW#reader_B004Y6ZBXW.
3. Taylor D. The pharmaceutical industry and the future of drug development. In: *Pharmaceuticals in the Environment*. Royal Society of Chemistry;2015:1–33. https://pubs.rsc.org/en/content/chapterhtml/2015/bk9781782621898-00001?isbn=978-1-78262-189-8.
4. Rappuoli R, Pizza M, Del Giudice G, et al. Vaccines, new opportunities for a new society. *Proc Natl Acad Sci U S A*. 2014;111(34):12288–12293. https://www.ncbi.nlm.nih.gov/pmc/articles/PMC4151714/pdf/pnas.201402981.pdf.
5. Rauch S, Jasny E, Schmidt KE, et al. New vaccine technologies to combat outbreak situations. *Front Immunol*. 2018;9:1963. https://www.ncbi.nlm.nih.gov/pmc/articles/PMC6156540/pdf/fimmu-09-01963.pdf.
6. Sahin U, Karikó K, Türeci Ö. mRNA-based therapeutics — developing a new class of drugs. *Nat Rev Drug Discov*. 2014;13(10):759–780. https://www.nature.com/articles/nrd4278.pdf.
7. Maude SL, Laetsch TW, Buechner J, et al. Tisagenlecleucel in children and young adults with B-cell lymphoblastic leukemia. *N Engl J Med*. 2018;378:439–448.
8. Whitehead E. *A Young Girl Beats Cancer with Immunotherapy*. https://www.cancerresearch.org/immunotherapy/stories/patients/emily-whitehead.
9. Emily Whitehead Foundation. *Stem Cell & Regenerative Medicine Action Award honoree*; 2020. https://www.youtube.com/watch?v=GW42FUhflkY.
10. Karikó K, Buckstein M, Ni H, et al. Suppression of RNA recognition by toll-like receptors: the impact of nucleoside modification and the evolutionary origin of RNA. *Immunity*. 2005;23(2):165–175.
11. Hilts PJ. *Protecting America's Health: The FDA, Business, and One Hundred Years of Regulation*. New York, USA: Alfred A Knopf Publishers; 2003.
12. Janssen WM. In: Food and Drug Law Institute, Levy MC, eds. *A Historical Perspective on Off-Label Medicine: From Regulation, Promotion, and the First Amendment to the Next Frontiers*. Washington, DC, USA: Off-Label Communications; 2008.
13. Janssen WF. *The Story of the Laws Behind the Labels*. FDA Consumer Magazine; 1981. https://www.fda.gov/media/116890/download.
14. Vargesson N. Thalidomide-induced teratogenesis: history and mechanisms. *Birth Defects Res C Embryo Today*. 2015;105(2):140–156.
15. Rägo L, Santo B. Drug regulation: history, present and future. In: van Boxtel CJ, Santo B, Edwards IR, eds. *Drug Benefits and Risks: International Textbook of Clinical Pharmacology*. revised 2nd ed; 2001. https://www.who.int/medicines/technical_briefing/tbs/Drug_Regulation_History_Present_Future.pdf.

16. *Pharmaceuticals and Medical Devices Agency*. https://www.pmda.go.jp/english/.
17. Wikipedia. *Pharmaceuticals and Medical Devices Agency*. https://en.wikipedia.org/wiki/Pharmaceuticals_and_Medical_Devices_Agency.
18. Singh K, Metha S. The clinical development process for a novel preventive vaccine: an overview. *J Postgrad Med*. 2016;62(1):4–11. https://www.ncbi.nlm.nih.gov/pmc/articles/PMC4944327/?report=printable.
19. FDA. *The Drug Development Process*; 2018. https://www.fda.gov/patients/learn-about-drug-and-device-approvals/drug-development-process.
20. EMA. *Human Medicines: Regulatory Information*; 2020. https://www.ema.europa.eu/en/human-medicines-regulatory-information.
21. AstraZeneca. *AZD1222 US Phase III Trial met Primary Efficacy Endpoint in Preventing COVID-19 at Interim Analysis*; 2021. https://www.astrazeneca.com/media-centre/press-releases/2021/astrazeneca-us-vaccine-trial-met-primary-endpoint.html.
22. Polack FP, Thomas SJ, Kitchin N, et al. Safety and efficacy of the BNT162b2 mRNA covid-19 vaccine. *N Engl J Med*. 2020;383(27):2603–2615. https://www.ncbi.nlm.nih.gov/pmc/articles/PMC7745181/pdf/NEJMoa2034577.pdf.

CHAPTER 4

COVID-19—The disease

1 Introduction

The COVID-19 pandemic, triggered by the severe acute respiratory syndrome coronavirus 2 (SARS-CoV-2) was and is a new challenge for our world. Its most important characteristic was not thousands of corpses on the streets and almost depopulated areas, as this was the case with historical epidemics. There are already websites that give a daily count of people who got infected with COVID-19, and of those who have died from it, until now, almost 5 million people by October 2021.[1,2] The number of infected patients and deaths are often used as evidence that COVID-19 is the worst epidemic/pandemic in human history. This is a short-sighted conclusion. Numbers alone are of limited value. There have never been so many people on our planet, and today's longevity is without parallels in previous human history. Unlike many other previous epidemics, the risk of dying from COVID-19 is relatively low. The prominent characteristic of the COVID-19 pandemic was and is the imminent blockade and clogging of the health structures, especially the acute hospitals. We are today used to a functioning healthcare system, structured in several hierarchical levels, and capable of managing accidents, infectious diseases, cancer, and many other challenges. As our society today is more complex, there is also an increasing number of perspectives from which to view the pandemic. Among the barriers to understanding the pandemic is the blurring of different perspectives, which makes it impossible to carefully work through the individual facets. Superficial blurring usually replaces a careful analysis with superficial assumptions, hackneyed prejudices, and flawed conclusions.

In any epidemic, the public interest initially focuses on the healthcare professional groups that care hands-on for the sick patients, i.e., doctors, nurses, physiotherapists, and others, and listens to their judgment. During the first year of the COVID-19 pandemic, many books were published that claimed to explain the essentials of the pandemic, often from authors already well known in their respective country; most, but not all of them were healthcare professionals.[3–6]

But their perspective was (and is) mostly short-sighted. Before we become infected with a pathogen, it must (1) exist at all; we must (2) encounter it where we are physically present, and it must catch us (3) in a state where our immune system does not immediately neutralize it when it tries to invade our body. The first three points include generally improved hygiene, nutrition, clothing, and much more in our modern developed world. Without modern means of transportation, the COVID-19 pandemic could not have spread so fast worldwide. Without modern communication, the pandemic would not be as much the focus of our attention as it is now. On the other side, point (2) also includes avoiding physical exposure through protective measures; reducing the risk of contagion through appropriate protective clothing; reducing the risk of contagion through lockdowns, curfews, restricting international travel; restricting local public transportation; ordering the wearing of masks in the public, in hospitals, and anywhere else; providing protective clothing to healthcare professionals; and more. But first and foremost there is the most important tool mankind has established against infectious diseases over the past two centuries: protective vaccination.

How things proceed from the moment you become infected depends on whether a hospital bed is available; what diagnostics and effective drugs, sufficient oxygen, and enough ventilators are available; the qualification and therapeutic experience of the local doctors and nurses, and much more. Since our society has become so complex, the representatives of the many healthcare professions and institutions are not only individuals but specifically in publications, interviews, and talking to the media they also represent their respective professions. We need to distinguish what the various experts say and what they mean. Experts can represent medical doctors; nurses; trade unions; institutions that teach public health and/or provide public health services; statisticians; epidemiologists; the World Health Organization (WHO); politicians; traditional and new media, including social media; and many more. Whenever one of them advocates strengthening this or that service, paying more attention to it, or whatever, this means also that the respective institution or professional group should be given greater weight compared to other institutions or professional groups.

2 DNA and RNA

Most living beings on our planet store their genetic information in very long double molecules that are composed of nucleotides and some other elements, entwined in a double helix. If for you it is sufficient just to memorize the expression "DNA", and if you do not have to take a biochemistry exam, then it is sufficient to associate "DNA" with a double (D) helix. If your biochemistry knowledge is more solid, then you can memorize that the backbone of DNA is made of deoxyribose. DNA stands for "**D**eoxyribo**n**ucleic **a**cid." DNA-stored information is used by the cells to assemble the various components of the respective cells and the entire body. The two DNA strands are composed of many units called nucleotides. Human development starts in the mother with the fertilization of a maternal egg cell by a sperm cell from the father. These two cells merge. The entire DNA of the later human is stored in this merged cell, which undergoes divisions, cellular differentiation, and growth, leading to a multicellular embryo, later a fetus, eventually a baby, and finally a human. DNA-stored information is responsible both for the development of the individual human from the first fused cell and for the continuous replacement of elements in the individual cell and the entirety of all cells. The code of the DNA is the order, in which the four nucleotides cytosine (C), guanine (G), adenine (A), and thymine (T) are lined up. If the genetic code is AACC, then this information will later be used to produce a different substance (protein) than ATAT or ACTA. Most higher organisms store their DNA inside the cell nucleus. Some more is also stored in other cell compartments. Both DNA strands store the same information, which is replicated when they separate and the work of translation begins.

In a library, the many books in themselves mean little. It takes people who read them and translate their content into action and/or new dreams, theories, stories, fairy tales, movies, reports, reviews, lies, or whatever else. Reading the books in the library corresponds in the cells to the genetic information being read and physically translated into cell components that replace or supplement existing structures and components. This is performed by complex machinery that uses the DNA-stored information as a blueprint.

One key element in the translation of genetic information is RNA. Like DNA, RNA is assembled as a chain of nucleotides. Again, there are four [C, G, A, U]. Three are identical to DNA [cytosine (C), guanine (G), adenine (A)]; the fourth one, U (uracil) is similar to thymine (T) in the DNA. For those who like to memorize: RNA stands for ribonucleic acid. Unlike double-helixed DNA, RNA is constructed as a single strand. It can fold, but then simply folds back on itself.

Higher organisms use DNA to store all genetic information, and RNA to reproduce individual proteins or other structural elements. Most DNA is stored in the nucleus of the cell, and cannot leave it. In contradistinction, RNA is everywhere, including the main body of the cell, the cytoplasm. Each individual RNA strand is much shorter than the long DNA strands. There are many different types of RNA. Messenger RNA (mRNA) carries information from the DNA in the nucleus to the sites of translation and element production in the cell main body. Essentially, the mRNA is a copy of genetic information stored in the DNA, but is now ready to be used in production. In these production processes, a limited part of the DNA is converted into RNA, respectively. Then the cell machinery translates this RNA into necessary physical new components. The translational apparatus that runs the translation from DNA-stored information down to the production of proteins are the ribosomes. Ribosomes are among the many sub-elements of cells. The ribosomes and associated molecules are also known as the translational apparatus.[7]

The discovery in 1953 of the structure of DNA as a twisted double helix by James Watson and Francis Crick was a milestone in the history of science and helped to establish today's molecular biology. It established the basis for our understanding of how genes control the construction processes in cells. Their discovery allowed an understanding of how genes are coded, how this translates into protein synthesis, how the pathways in these processes can go wrong, and how flawed pathways can be corrected. Eventually, this led to recombinant DNA research, genetic engineering, monoclonal antibodies, and the development and production of further biologics on which today's multi-billion dollar life science industry is founded. Of course, Watson and Crick did not simply stumble over the DNA structure by chance or on their own. Generations of earlier researchers had contributed earlier understandings and had made important but yet unconnected findings of the composition of DNA. Watson and Crick put these findings together into a coherent theory of genetic transfer. They published their findings in a one-page paper, titled "A Structure for Deoxyribose Nucleic Acid," in the British journal *Nature* in April 1953, illustrated with a schematic drawing of the double helix, done by Crick's wife. They tossed a coin to decide the order in which they were listed as authors. Among the novel features of considerable biological interest they described the pairing of the bases on the inside of the two DNA backbones: A with T and C with G. This pairing rule suggested a copying mechanism for DNA: Given the sequence of the bases in one strand, the other was automatically determined. If got separated, each could serve as a template for a complimentary new chain. Watson and Crick developed their ideas about genetic replication in the second article in Nature, published in May 1953.[8–11]

The understanding that there are two types of genetic coding, i.e., DNA and RNA, emerged in the late 1950s and the 1960s. The concept emerged that DNA leads to the formation of RNA, which in turn leads to the synthesis of proteins. During the early 1960s, DNA and RNA were studied more and more. The concept of many different types of RNA emerged. One of them is the messenger RNA (mRNA). There was a multitude of obstacles that had to be overcome step by step. One was the short-lived nature of RNA which made its biochemical isolation difficult. Eventually, this problem was overcome in the 1960s by using special cells (reticulocytes) from vertebrates that produce large quantities of mRNA.[12]

RNA therapeutics is a rapidly expanding category of drugs that might change the standard of care for many diseases in the near future. First RNA drugs have been approved, many more are in development. We might be in the middle of a new therapeutic revolution, comparable to the beginning of recombinant protein technology decades ago in Silicon Valley.[13] Many patients participate in large RNA clinical studies to fight cancer, renal and hepatic disorders, ocular and cardiovascular disease. Initially, RNA was thought to be a poor choice for a therapeutic agent, given its relatively short half-life. However, with improvements in stabilization, much of this skepticism has been overcome.[14] Naked, single-stranded RNA is degraded fast in the cytoplasm, can activate the immune system, and is too large and negatively charged to just cross the cell membrane. For therapeutic use, it must therefore be equipped with additional means of cellular entry. RNA delivery has centered on the design of delivery methods and materials that aim to transport RNA drugs to the site of interest. RNA delivery can be done by many different delivery methods. One way is engineering adeno-associated viruses to carry nucleic acid cargo. Nanoparticle encapsulation protects nucleic acids from degradation and can aid in cellular uptake and endosomal escape.[15] The biological understanding of RNA has evolved from simply an intermediate between DNA and protein to a dynamic and versatile class of molecules that regulate the functions of genes and cells in all living organisms. This has led to the emergence of numerous types of RNA-based therapeutics that broaden the range of potential drug targets beyond the scope of existing pharmacological. Messenger RNA (mRNA) holds not only the potential to revolutionize the development of vaccines, but also of protein replacement therapies, the treatment of genetic diseases, the treatment of cancer, and many more.[16]

The ability to understand and work with the genetic code emerged from the convergence of at least three areas. New methods to generate synthetic RNA were developed to serve as artificial mRNA. Systems were developed that were able to translate synthetic mRNA into protein. It also emerged that the genetic code uses three-letter "words" (codons).[11,12] Proteins are found in every cell and usually make up more than half of their dry weight; the largest component of human tissue is water. Proteins serve as molecular tools and perform different tasks, including cell movements, transporting metabolites, pumping ions, catalyzing chemical reactions, or recognizing signal substances. Muscles, heart, brain, skin, and hair also consist mainly of proteins. Codons are the three-letter-"words" (three-nucleotide sequences of DNA or RNA) that correspond, respectively, to a specific amino acid. Beyond their role as residues in proteins, amino acids participate in several processes such as neurotransmitter transport and biosynthesis. Proteins consist of amino acids. The genetic code is the relationship between the order of letters in the DNA or RNA and the resulting sequence of amino acids. There are 64 different codons, of which 61 specify amino acids. The remaining three are stop signals.[17] Today, our understanding of the genetic code permits the prediction of the amino sequence of the protein products of the tens of thousands of genes whose sequences are being determined in genome studies.[12]

Over time it was also recognized that retroviruses have a single-stranded RNA genome and that they can replicate via an intermediate DNA, i.e., that they could reverse the usual DNA-to-RNA transcription pathway. Some retroviruses can cause diseases, including several that are associated with cancer. The human immunodeficiency virus (HIV), another retrovirus, is the cause of acquired immunodeficiency syndrome (AIDS). The enzyme involved in the reverse transcription from RNA to DNA (reverse transcriptase) was and is widely used as a tool for the analysis of RNA in the laboratory.[12] Over time, many more types of RNA were identified in the genetically coordinated processes of construction and breakdown in the cell.

In the context of the COVID-19 pandemic, understanding DNA vs RNA is essential from at least two aspects. First, many viruses, including the coronaviruses, store their genetic information in the form of RNA. Second, the most modern COVID-19 vaccines are based on messenger RNA (mRNA), which has many advantages over traditional vaccine development strategies.

The global research process had initially resulted in the discovery of the genetic code and the decoding of the human genome. With the discovery and growing knowledge of RNA, a type of molecule that has only been known for half a century, the basis for a whole new type of industry arose as part of the further gradual transition from the traditional "Big Pharma" to a new high-tech advanced life science industry, which has played a key role in the development of the COVID-19 vaccines. In the following, we show how the discovery of the key elements in the understanding of DNA and RNA and the growing understanding of the processes in the individual cells and the entirety of the respective human and animal organisms rub against the existing institutions, entrenched doctrines, and outdated political structures. The development of vaccines was on one side a straightforward technical process on the background of the continuously improving industrial development and production of medicines and vaccines. These medicines and vaccines are becoming more and more an integrated part of the entire human society with all its open and hidden contradictions and conflicts. On the other hand, this development was and is based on the tortuous path of scientific progress, which has to overcome many dead ends, misconceptions, and setbacks. In this book, we primarily focus on the interfaces between powerful institutions, learned societies, associations, and illusions that influenced the path the COVID-19 vaccines were conceptualized, produced, and finally approved. How they had to go from initial concepts and ideas into a physical vaccination campaign that already restricts the pandemic, will continue to drive this forward in the near future and will defeat it eventually. All this also rubs against existing institutions, entrenched doctrines, outdated political structures, and flawed illusions that mankind will need to overcome in the immediate future.

3 Severe acute respiratory syndrome coronavirus 2 (SARS-CoV-2) and COVID-19 disease

Coronaviruses were first described in the 1960s. They carry their genetic information as a single-stranded RNA, which is embedded in an envelope. The virus has four essential structural proteins which include the spike (S) glycoprotein, matrix (M) protein, nucleocapsid (N) protein, and small envelope (E) protein. The spike proteins have contributed to the naming of this group of viruses. Corona means "crown" in Latin, due to the virus' appearance under electron microscopy. The virus' trick to infect cells is that it binds with the structural elements on the spikes (on the outside of the envelope) to specific receptors on the surface of targeted cells unless the virus is neutralized before attempting to do so. The receptor on the outside of the host cell is called "angiotensin-converting enzyme 2" (ACE2). Unless you prepare for a biochemistry exam, just memorizing "ACE2" will do. After the binding of the spike (S) protein with the ACE2 receptor, the invasion of the host cell begins. The viral RNA is released into the host cell and gets translated. When the newly synthesized viruses are released from the infected cells, this activates the host's immune system. If the host is lucky, white blood cells attack and destroy the viruses. Furthermore, inflammatory interferons are produced to help the destruction and removal of the viral antigens. Also other parts of the immune system kick in. In most persons, the initial inflammation is time-limited and ends with immunity to future COVID-19 infections. However, if the immune system goes into overdrive and the body cannot control the inflammatory cascades, then there is a danger to the patient's life.[18–20]

Coronaviruses spread among humans, mammals, and birds. Middle east respiratory syndrome (MERS) and severe acute respiratory syndrome (SARS) were two global epidemics in the recent past that were caused by two other types of coronaviruses.[21] At present, we know seven coronaviruses that infect humans, causing diseases ranging from a mild form or common cold to severe and/or fatal infections.

The severe acute respiratory syndrome coronavirus 2 (SARS-CoV-2) was first observed and reported in Wuhan, China, in December 2019. Li Wenliang, a Chinese ophthalmologist, raised awareness of early infections. He reported about several suspected severe acute respiratory syndrome (SARS) that had been brought to the hospital after contact with the Huanan Seafood Market. He did this in a chat room and asked for this information to be kept confidential, but his observations were circulated broadly. He was interrogated by the local police and was issued a formal written warning. He was made to sign a letter of admonition promising not to do it again. The police warned him that any recalcitrant behavior would result in prosecution. He returned to work at the hospital, contracted the virus, published his experience in the police station with the letter of admonition on social media, and died at the end of his COVID-19 infection.[22] A lockdown in Wuhan and other cities in the surrounding province was organized, but during its preparation, thousands fled to other parts of China. Eventually, China managed with its strict lockdown policy to largely contain the spread of the virus in China. But the virus had already spread to other parts of the world. The World Health Organization (WHO) declared a Public Health Emergency of International Concern in January 2020, and a pandemic in March 2020. Since then, variants of the virus have emerged that have become dominant in many countries. So far, almost 5 million deaths have been confirmed.[1,2,23–25]

The clinical manifestations of COVID-19 range from no symptoms at all to cough, fever, fatigue, breathing difficulties, sore throat, loss of smell and taste, pneumonia, and multisystem failure, which can lead to death. The beginning of symptoms can be soon after infection, or up to 2 weeks later. In most persons, it causes only mild symptoms or none at all. It is mainly transmitted by inhaled droplets. The body's immune response leads either to an unproblematic defeat of the virus, or to a highly inflammatory response with immune dysregulation, a cytokine storm, acute lung injury, acute respiratory distress syndrome (ARDS), multiple-organ system failure, and death. After the first infection, with or without few or many symptoms, organs can be damaged long-term, including fatigue, memory loss, and more. Serious diseases and an impaired immune system carry the highest risk of severe disease and death. However, older age is the greatest risk factor for a severe course.[26–28]

The natural reservoir for coronaviruses is bats. It appears that the severe acute respiratory syndrome coronavirus 2 (SARS-CoV-2) originated in a bat population in China. Furthermore, highly homologous SARS-CoV-2 types were found in smuggled Malayan pangolins. The large-scale explosion of the COVID-19 is conventionally traced back to the wet markets of Wuhan, China. In Southern China, there is the habit of consuming a wide range of exotic animals including different species of bats as food or for traditional Chinese medicine. The precise avenue how SARS-CoV-2 went from bats to humans, with or without an intermediate step of pangolins or any other animal, is still not exactly known and will certainly keep many researchers busy for the next years. We will still have to understand why previous similar coronavirus infections failed to develop the pandemic dimensions of COVID-19; there might have been an intermediate host in the transmission from bat to humans; and more.[28–31]

The speed at which the virus spread from person to person took the world by surprise. The WHO organized an international fact-finding on the origins of the pandemic, but handled it more like an international diplomatic inquiry, and not like an independent scientific investigation.[31,32] Seventeen experts with the WHO teamed up with 17 Chinese scientists to assess four potential scenarios for the origins of the coronavirus. But Chinese scientists refused to share raw data. Disagreements over patient records and other issues were so tense that they sometimes erupted into shouts among the typically mild-mannered scientists on both sides. China's continued resistance to revealing information about the early days of the coronavirus outbreak was and is certainly not helpful in understanding the origins of the pandemic.[32,33] Nevertheless, the team concluded that two leading scenarios are the transmission of the virus to people either directly from bats or, more likely, via an intermediate animal. A third possibility is the virus got to people through contaminated frozen food products, which the team considers less likely but says merits further investigation. The last scenario—that the virus began spreading among people following a lab accident—is "extremely unlikely," the researchers wrote. The WHO report tallies where the evidence currently points: The virus, called SARS-CoV-2, probably jumped to people from bats through another animal; it likely did not come from a lab.[34] But officials can not yet prove any scenario or rule it out. Questions about just how much access to potential evidence the international team of experts had on their 28-day trip to Wuhan, China, in January and February 2021 has cast a shadow on the findings. In a joint statement on March 30, 14 countries including the United States expressed concern that the WHO team was delayed and did not have access to original data and samples from people and animals.[35] After the WHO report, the theory that the COVID-19 pandemic was caused by a laboratory accident from the Wuhan Institute of Virology, which is very near the Wuhan wet market, has again been discussed repeatedly. China's unwillingness to share all relevant data and facts has meanwhile been publicly criticized by the WHO Director-General, Tedros Adhamon Ghebreyesus, and others.[36] On the other hand, apart from the repeated utterance of this allegation, no relevant new facts have emerged.

4 Modern society and public health

Prevention, diagnosis, and treatment of diseases are today part of a complex interaction of which multiple vaccinations are just a tiny part. For vaccinations to be carried out, the vaccines must be developed, approved, produced, distributed, stored, and finally administered by qualified personnel. Everything must be well documented. The institutions and professions involved are the doctors and nurses; the pharmaceutical industry, which develops and produces drugs and vaccines; regulatory authorities; pharmacies; and lawyers in case anything goes wrong. If someone sells counterfeit drugs or vaccines, customs and/or the police intervene. In a broader sense, there are also the media that report on the health system, public education in which young people receive minimal knowledge about illness and health (or not); public administration; the government, which in its own opinion always does everything right; the opposition, in whose opinion the government does everything wrong (if there is an opposition); the universities, where knowledge is processed and further developed in interaction with industry, and many more.

With few exceptions, persons undergo multiple vaccinations from the first days of life, continued, repeated, and boosted in different schedules. Together with the introduction of clean water and sanitation, vaccination has probably made the greatest contribution to global human health. Child mortality declined continuously and progressively already before the development and deployment of vaccines, largely due to a reduction in mortality from infectious diseases due to better housing, nutrition, sanitation clothing, and more. But there is no doubt that vaccination made an enormous contribution to human health, especially in the developing world. Mortality from smallpox and measles was massive in the pre-vaccination period with up to a half of the population dying from smallpox during epidemics, and measles only a little less lethal in susceptible populations.[37-39] In adolescents, boosters are required for pertussis, meningococcus, diphtheria, tetanus, and influenza. Furthermore, female adolescents are vaccinated to prevent infection with the human papillomavirus. Also, adults should repeat booster injections for some diseases such as tetanus. Furthermore, new and better vaccines are developed today also for elderly persons whose aged immune system might need more potent vaccines.[40,41] Global coverage of vaccination against many childhood infectious diseases has improved considerably since the creation of the WHO's Expanded Program of Immunization in 1974 and the Global Alliance for Vaccination and Immunization (GAVI Alliance) in 2000, today GAVI, the Vaccine Alliance.[42-44]

When in 2006 mainstream media first took notice of human fatalities caused by the deadly bird flu, public access to the latest genetic sequences of highly pathogenic avian influenza was limited and often restricted due to the hesitancy by affected countries to share their information through traditional public domain archives. Public domain archives, where access and use of data take place anonymously, did not offer protection of intellectual property rights, or incentives to share such data, such as recognizing the submission of the data. In 2008, a new platform was created under the name "Global initiative on sharing avian influenza data" (GISAID).[45,46]

It has played an essential role in the sharing of data among the WHO Collaborating Centers and National Influenza Centers for the regular influenza vaccine virus recommendations by the WHO Global Influenza Surveillance and Response System (GISRS), a global network of laboratories that has for purpose to monitor the spread of influenza with the aim to provide the WHO with influenza control information. Established in 1952, GISRS is coordinated by WHO and endorsed by national governments.[47,48] In 2010 Germany became the official host of GISAID.[45,46] GISAID now plays a major role in uploading and comparing new genetic SARS-CoV-2 variations strands worldwide.

In the COVID-19 pandemic, genomic sequencing was possible for the first time in real-time.[49] The virus' genome sequence was released on an open-access Virological website early in January 2020.[50,51] Thereafter, the China CDC released SARS-CoV-2 genome sequences (with associated epidemiological data) on the public access database GISAID.[45,46,52,53]

5 Traditional avenues of prevention

When the first COVID-19 cases were reported and the Chinese authorities had abandoned their initial reflex to deny everything, China acted quickly and decisively. A lockdown was declared in Wuhan and the province, resulting in an initial containment of the pandemic. The drastic measures included countrywide lockdowns, face mask mandates, massive control by testing, and quarantine for those that had been in contact with persons infected with COVID-19. Discussions about the outbreak were censured since the beginning of the pandemic. Reporting and criticism about the crisis were censored and countered. Several citizen journalists were arrested. The official response to the outbreak was portrayed in a positive light. Eventually, China managed to bring the outbreak under control.[23] China is a police state where the individual does not enjoy any rights. Methods like those in China are unimaginable in most western countries. By late February 2020, the pandemic had been brought under control in most Chinese provinces.

Most countries worldwide were surprised by the speed of the pandemic. Most first cases went unnoticed. As the extent of the infection began to be realized, it emerged that most countries were poorly prepared for a pandemic. There was a lack of face masks, protective clothing, ventilators, and much more.

General public health measures comprise everyday preventive actions include wearing masks in public, staying six feet away from others, avoiding poorly ventilated spaces and crowds, washing hands with water and soap more frequently, covering coughs and sneezes, cleaning regularly touched surfaces every day, monitoring the own health regularly for fever, toxins, breathing difficulty, or other COVID-19 symptoms, staying home when sick, avoiding crowds and unnecessary traveling, and try to increases the body's immunity by adopting healthy habits, such as a balanced diet, adequate rest, oral hygiene, regular exercise, and avoiding excessive fatigue and stress.[54]

Many countries and regions imposed quarantines, entry bans, or other restrictions for citizens of or recent travelers to the most affected areas. Other countries and regions imposed global restrictions on foreign countries and territories, including preventing their citizens from traveling overseas. Travel restrictions have probably reduced the spread of the virus. However, as they were implemented after the pandemic had already spread around the globe, their contribution was not decisive. In most countries, they contributed only to a modest reduction in the number of persons infected. The travel restrictions caused massive economic harm to the tourism industry and the aviation industry through lost income and social harm to persons who were prevented from traveling for family matters or other reasons. Travel bans may be most effective for isolated locations, such as small nations. Researchers concluded that travel restrictions are probably most useful in the early and late phase of an epidemic; the restrictions of travel from Wuhan, China, came unfortunately probably too late.[55,56]

As other countries battled months-long lockdowns and hospital systems on the brink of collapse, New Zealand imposed a five-week nationwide lockdown before returning to something resembling normality. Borders had been closed to foreigners for more than a year, but music festivals and weddings have gone ahead. The country reported around 2500 cases of COVID-19 and 26 deaths. It was comparable in other parts of Asia-Pacific. When the Australian think tank Lowy Institute scored more than 100 countries on their COVID-19 performance during the first year of the pandemic, it judged that Asia-Pacific had, on average, been the most successful region in the world at containing the pandemic. The two worst regions, according to the Lowy Institute, were Europe and the Americas – and leaders in the United States and the United Kingdom, in particular, were seen as catastrophic failures in their handling of the pandemic. While the UK and the US have increasingly vaccinated their population, New Zealand, Thailand, Taiwan, South Korea, and Japan, who had all been relatively successful at preventing large-scale outbreaks, had each vaccinated only a few percent of their populations. The situation differs for each country, but one reason was and is that they did not sign agreements with manufacturers for vaccines as early as others. Countries in Asia-Pacific did not have the same sense of urgency when they signed contracts. The UK and the US bettered on the efficacy of future vaccination before other countries. Countries in Asia-Pacific initially handled the pandemic well. They avoided paying the huge economic and social costs that other nations overseas have paid. The economies of South Korea, New Zealand, and Australia were hit, but not as bad as other countries.[57,58] Nevertheless, on a strategic level, there is no way around effective vaccination.

The US had about one-fifth of the worldwide COVID-19 cases and deaths. COVID-19 was in 2020 the third-leading cause of death in the US, behind heart disease and cancer. Life expectancy dropped by roughly 3 years for Hispanic Americans and African Americans, and roughly 1 year for white Americans from 2019 to 2020.[59] The Trump administrations initially held regular coronavirus briefings when the pandemic began. However, these briefings became inconsistent and faded out in May 2020, despite rising cases at the time. President Trump admitted in March 2020, that the virus was "bad" because it was so contagious, but then, frequently throughout the pandemic, he changed his tune. His message was consistently inconsistent, and at times even became dangerous, when, e.g., he praised the efficacy of hydroxychloroquine (which turned out wrong) and told people not to be afraid of COVID after he had recovered from the virus,[60] thanks to excellent medical care far above the level many other US citizens enjoyed.[61] In October 2020, the prestigious New England Journal of Medicine published an editorial, signed by all 34 editors, in which they condemned the Trump administration's handling of the COVID-19 pandemic, describing it as dangerously incompetent and that it had turned a crisis into a tragedy.[62] It was the first time that the NEJM has ever supported or condemned a presidential candidate and only three other times in history has an editorial been signed by all the editors.[63]

In summary, traditional protective measures can have a positive influence on the course of a pandemic to a certain extent. Being an island helps initially. Nevertheless, strategically there is no way around against a viral infectious disease but to prevent it through effective vaccination.

6 Vaccines

Vaccines are biological preparations that help the body to build up immunity against infectious diseases. They stimulate the immune system to recognize the respective harmful agent as a threat, destroy it, and to further recognize and destroy the same agent if it is exposed to it in the future again.

The human immune system matures during the first months and years of life. Initially, there is some protection against infectious diseases the mother has gone through many years earlier. This protection is provided by antibodies transferred from the mother to the child through the placenta before birth, and in the mother's milk after birth. Gradually, this initial protection is extended and replaced by the child's maturing own innate and adaptive immune systems.

The Trump administration failed fundamentally in organizing the traditional measures that could have slowed considerably the spread of COVID-19 in the US. On the other hand, the same administration took a key strategic first step toward a way out of the COVID-19 pandemic in the medium and long term. It became the first light at the end of the tunnel. This step was named "Operation Warp Speed" (OWS). A warp drive is a fictional spacecraft propulsion system in science fiction, most notably Star Trek, which began as a TV series in 1966 and turned into many movies, books, magazines, comics, action figures, model toys, and video games, representing an entire fictional universe set in the future. A spacecraft equipped with a warp drive may travel faster than light.[64] OWS thus stood for a vision that was supposed to accomplish the almost impossible in a short time. But OWS also reflected the American vision that technical solutions for complex challenges are possible. And here we are at the parallel to World War II, which was ultimately brought to an end by the successful Manhattan Project faster than would have been possible only with conventional warfare.

OWS was officially announced in May 2020 by the US government as a public-private partnership to facilitate and accelerate the development, manufacturing, and distribution of COVID-19 vaccines, therapeutics, and diagnostics.[65] It aimed from the beginning to develop, produce and deliver 300 million doses of safe and effective vaccines with a January 2021 target.[66–68] It was headed initially by Moncef Slaoui, a researcher who was the former head of the vaccines department at GlaxoSmithKline (GSK), a "big pharma" company for which he had worked for 30 years, retiring in 2017. After Moncef Slaoui, OWS was headed by David A. Kessler, a US pediatrician, attorney, author, and administrator (both academic and governmental). He had been the commissioner of the FDA from 1990 to 1997. February 2021, OWS was transferred into the responsibilities of the White House COVID-19 Response Team, with David A. Kessler as Chief Science Officer. OWS promoted mass production of multiple vaccines, and different types of vaccine technologies, based on preliminary evidence, allowing for faster distribution if clinical trials would confirm one of the vaccines is safe and effective. The plan accepted from the very beginning that some vaccines would not prove safe or effective. This made the program more costly than typical vaccine development but offered the potential to result in the availability of a viable vaccine in an accelerated manner.[65,69] Initially, a budget of $US 10 billion was assigned, with additional funds available.[66,70] In the end, it had invested $US 18 billion.[67] OWS included components of the US Department of Health and Human Services (DHHS), the Centers for Disease Control and Prevention (CDC), the FDA, the National Institutes of Health (NIH), the Biomedical Advanced Research and Development Authority (BARDA); the US Department of Defense; private firms; and several further US federal agencies.

The COVID-19 pandemic has thus resulted in the bizarre situation that the same federal US administration on one side almost completely failed on the tactical level of conservative restraint of the pandemic and left many key decisions in the hands of the individual US states, which reacted in very different ways to the pandemic. However, it also resulted in a good decision on the strategic level that has the potential in itself, the pandemic soon to overcome.

Operation Warp Speed has been of tremendous aid in the manufacturing and development of the BNT162b2 vaccine without cutting corners to jeopardize its safety and efficacy.[71–75] The fast approvals of three vaccines by the FDA represent a public health breakthrough, providing the first protective measures against the largest global pandemic to strike in over 100 years, and were the first fruit of a vaccine development process akin in scope and urgency to the famed Manhattan Project. The Manhattan Project had managed the development of the first atomic bombs during World War II.[76] The comparison between World War II and the COVID-19 pandemic is not far-fetched. The obstacles that had to be overcome were different, but this comparison is certainly legitimate if we consider the dimension of these challenges.

Three COVID-19 vaccines have now received FDA Emergency Use Authorization,[77] one has now full FDA approval,[78] four have been approved by the European Medicines Agency (EMA),[79,80] and three by the UK MHRA.[81]

When US President Trump was succeeded by Joe Biden, more than sufficient vaccines had already been ordered from the different vaccine manufacturers. And, equally important, they had already been paid for. With the new president, there were now sufficient vaccines available in the US to initiate immediately with an efficient vaccination campaign. The European Union (EU) had a slow vaccine rollout. The EMA was rather slow to approve the first vaccines. Then, the manufacturer AstraZeneca could not deliver as fast as it had expected. But the contract between the EU and AstraZeneca was not as binding as, e.g., the contract between the UK and the EU. The EU took AstraZeneca to court at the European Court of Justice, which, however, did not accelerate the delivery of the vaccines. Then, several EU countries halted their use of the AstraZeneca vaccine due to safety concerns, a move that baffled health experts and did not contribute to the public trust in the vaccination campaign. The contracts between the manufacturers were concluded late and were not binding. As so often the case with the EU, too many actors were involved in decision-making. In the end, it was virtually impossible to pinpoint who was responsible for what. The EU Commission conducted the negotiations with the producers, but under the control of a committee of representatives from the member states, recreating the coordination problems that centralized procurement was meant to avoid. The EU still has a good base for scientific research and the capacity to produce new high-tech drugs and vaccines fast and at scale. But the EU's structure is not suited to fast executive action. Its decision-making mechanisms are complicated and make accountability difficult.[82,83]

COVID-19 is mainly transmitted via the respiratory route when people inhale droplets and particles that infected people release as they breathe, talk, cough, sneeze, or sing.[26,27] The COVID-19 virus gains entry into human cells with a special key at the top of the spikes that have given the virus the name "corona" (crown). Two of the three spike proteins at the tip of the spike are involved, called the subunits S1 and S2. This key binds to a specific enzyme, whose name you do not need to learn by heart Of course, an infection occurs only if the viruses can reach their target cells unhindered. If the attacked person has antibodies that bind to the spike protein, they prevent its attachment to the host cell and with this neutralize the virus. Furthermore, in addition to antibodies, natural infection or successful immunization trigger immune responses that recognize and help to destroy those viruses they have recognized as hostile and potentially harmful.[84]

All vaccines approved by the FDA, the EMA, and the MHRA are given intramuscularly. They contain the genetic information for the composition of the proteins on the spikes on the outside of the C virus. The vaccines make the vaccinated person's muscular cells produce the spike protein that in unprotected humans would work as the key that allows the virus to enter the host cells. This spike protein, which is now produced by the own body, is presented to the vaccinated person's immune system, which now recognizes it as potentially harmful. This induces protective immunity by priming immune cells (T cells) that later recognize viruses that carry the spike protein and organize a neutralizing antibody response. The antibodies that are triggered by all COVID-19 vaccines currently approved by FDA, EMA, and MHRA trigger antibodies known as virus-neutralizing antibodies (VNAs). From the moment the body has learned to recognize the spike protein (which has been produced by the body's own cells!) attacks the viral spike protein and thus prevent the virus from invading the body. Virus neutralizing antibodies (VNAs) are not the only arm of the immune response required for protection, but they comprise an essential component.[68]

Messenger RNA (mRNA) was discovered several years later and has since then been the subject of consistent basic and applied research for various diseases. mRNA does not need to enter into the nucleus to be functional; it works once it has reached the cytoplasm. Furthermore, mRNA-based therapeutics does not integrate into the cell's genome. For most pharmaceutical applications, mRNA is only transiently active and is thereafter degraded via physiological metabolic pathways. The production of mRNA is also relatively simple and inexpensive. The development of mRNA-based therapeutics has triggered broad interest.[85,86]

Altogether, decades of research on the human genome project, of RNA, and delivery methods of RNA therapeutics have all been used in the development of COVID-19 vaccines. We can in this book not discuss all technical aspects that the RNA research had to go through to use RNA and mRNA for vaccinations and other purposes. However, in the following, a short discussion of some of the discoveries that made the development of RNA vaccines possible.

In 2005 it was learned how to modify mRNA so that it could evade immune detection and boost protein production.

Katalin Karikó, a Hungarian researcher who had immigrated to the US, attempted to harness the power of mRNA to fight disease. She was aware that the body relies on millions of tiny proteins to keep itself alive and healthy, and that it uses mRNA to tell cells which proteins to produce. If it was possible to design a specific mRNA, this would make it possible to hijack the process of protein creation and production. Synthetic RNA was known to be vulnerable to the body's natural defenses and would be destroyed before reaching its target cells. Furthermore, the resulting biological havoc might stir up an immune response that could make the therapy a health risk. As discussed above, every strand of mRNA is made up of four molecular letters called nucleosides. In an altered synthetic one of these four letters was throwing everything off by signaling the immune system. Karikó and Weissman subbed this one letter (nucleoside) by a slightly modified version. With this, they created a hybrid mRNA that could sneak its way into cells without alerting the body's defenses. They proposed to replace a natural component of RNA with the modified, non-natural RNA nucleobase N1-methylpseudouridine (m1Ψ) to enhance immune evasion and protein production. When they published their idea, few took it seriously. The second key step was that suitable formulations are required to protect in vitro transcribed (IVT) mRNA against degradation outside of the cell and to facilitate its entry into cells. Once you have identified the right complexing agents, they are able to protect the mRNA from degradation and also act as facilitators for its cellular uptake. These thoughts were not developed specifically in the process of vaccine development. On the contrary, they emerged in the quest to develop anticancer medications that are able to target specifically and personally a patient's cancer cells and overcome its defensive strategies. There were many more complex steps involved, but here we just try to outline the basics. An additional advantage of mRNA-based therapeutics is that they do not integrate into the genome which is stored at the cell's nucleus. Therefore, they do not pose the risk of causing mutations. Also, synthetic mRNA is active only for a short period and is then completely degraded by the cell's and the body's metabolic pathways. And the production of in vitro transcribed (IVT) mRNA is relatively simple and inexpensive.[84–88]

Overall, we see a scenario of open scientific exchange on the one hand. Chinese scientists published the entire genome of the SARS-CoV-2 virus on the internet very early, making it accessible to all other researchers worldwide. On the other hand, there is the implementation of scientific findings in patents and company foundations. There is now a large, vibrant scene of companies researching drugs and vaccines based on RNA technology.[89,90] This scene is very competitive. Anyone who cannot assert themselves as a company goes under and/or is swallowed up. These companies are no longer just the big old "Big Pharma" companies. On the contrary. The companies BioNTech, Moderna, and Curevac are relatively small and new. If large companies are flexible enough, they can successfully form an alliance with one or more small companies and achieve great success together, see for example the alliance between Pfizer and BioNTech. Thirdly, there are the political strategies that are reflected in various directions of state supervision. The Chinese government was not helpful and transparent when the origin of the SARS-CoV-2 virus in China was investigated. This has so far contributed massively to massively diminishing the credibility of everything that officially comes from Chinese sources. There are many more factors.

The three FDA-approved vaccines are two mRNA vaccines from Pfizer-BioNTech and Moderna, given as two injections about a month apart, and one from Jannsen to be administered as a single dose injection.[77] These approvals represent a fundamental public health breakthrough and were the first fruit of a vaccine development process akin in scope and urgency to the famed Manhattan Project.[76] The four EMA-approved vaccines are the two mRNA vaccines from Pfizer-BioNTech and Moderna, the vaccine from AstraZeneca, and the one from Janssen. Both the AstraZeneca and the Janssen vaccines employ a common cold-causing virus that is genetically engineered so that it cannot replicate in the vaccinated person. Into this genetically engineered virus, the encoding sequence of the SARS-CoV-2S protein has been inserted. Similar to the mRNA approach, the viral vector approach uses the host cellular machinery for transcription of the SARS-CoV-2S protein gene to mRNA and then translation to the SARS-CoV-2S protein. The three MHRA-approved vaccines are of BioNTech, Moderna, and AstraZeneca.[77] Thus, all vaccines approved by the FDA, the EMA, and the MHRA let the vaccinated person produce the crucial protein of the COVID-19 virus and then let it go from the producing cells into the whole body.[91] This spread of the virus protein occurs when it exits the originating cells and enters extracellular space and with this the blood circulation, where it encounters the body's own immune system. The RNA used to produce the required virus protein is broken down and disposed of.

Despite the urgency of the entire vaccine development programs, the development of SARS-CoV-2 vaccines went through the same steps as all other vaccines during their development process, from target identification over the design of candidate vaccines, and finally through the required human clinical trials. Initial human studies were performed on dose-finding, safety, and the development of antibodies (measures of immunogenicity). The phase III trials were double-blind, randomized, and placebo-controlled. Furthermore, a minimum of 2 months of safety data was required for consideration for emergency use authorization (EUA). Operation Warp Speed (OWS) coordinated across different studies that collected similar data; endpoints were harmonized, and standardized assays were used to measure immune response and other biologic feedback data. OWS also monitored enrollment to assure that each trial had sufficient participants of elderly persons and ethnic minorities, including Hispanic people, black people, those with medical co-morbidities, and more.

7 Diagnostics

There have been considerable advances in in vitro diagnostic (IVD) assays for COVID-19 infection. The main IVD assays use real-time reverse-transcriptase polymerase chain reaction "[(real time PCR; RT-PCR) that takes a few hours (it is sufficient to memorize 'real-time PCR')]". Meanwhile, the assay duration has been shortened to 45 min, and another molecular assay has decreased the assay duration to just 5 min. Most molecular tests have been approved by the United States Food and Drug Administration (FDA) under emergency use authorization (EUA) and are Conformité Européenne (CE) marked. Furthermore, many serology immunoassays have also been developed that allow an assessment of the respective person has gone through a COVID-19 infection.[92] As of August 2021, the FDA has issued 255 individual emergency use authorizations (EUAs)[93] for in vitro molecular testing for COVID-19,[94] 33 antigen diagnostic tests,[95] 87 tests for Serology and Other Adaptive Immune Response Tests for SARS-CoV-2,[96] and 4 tests for in vitro diagnostics for management of COVID-19 patients.[97]

Also the EU has released an impressive list of approved COVID-19 rapid antigen tests and a common standardized set of data to be included in COVID-19 test result certificates.[98]

8 Treatment

For the optimal functioning of vital organs, existing therapeutic procedures for patient diagnosis and recovery focus in the first place on symptomatic treatment and subsequent supportive care. The primary interventions include bed rest, supportive care, adequate calories, water consumption, water balances and homeostasis, psychotherapy, oxygen therapy. Vitamin C and probiotics can also be used to improve the immune response. The very first step is to ensure that the infection is not further transmitted to family members, patients, and healthcare staff by proper insulation. Mild sickness should be treated in the home with appropriate advice on threat signs and strictly follow health practitioner recommendations. COVID-19 management furthermore includes treating cough and fever along with hydration and nutrition monitoring. In hospitals, oxygen is recommended in hypoxic patients via a facial mask, a nasal cannula, or non-invasive ventilation. Mechanical ventilation and extracorporeal membrane oxygen support may be needed if necessary, and renal replacement therapy might be required.[54,99]

Once somebody is infected with COVID-19, there is no elegant way just to interrupt the infection and restore the patient's health. As listed below in Table 1, as of October 4, 2021, the FDA has issued 11 Emergency Use Authorisations (EUAs) for drugs and biological products.[100] Many more medications are currently under clinical trials for the treatment of COVID-19.[54]

Remdesivir directly inhibits the RNA-dependent polymerase of nCoV, thus declared the most proficient drug molecule available for COVID-19.[54,101]

Chloroquine, an antimalarial drug, was praised by former US President Trump as a remedy against COVID-19. However, current CDC guidelines warn that chloroquine and hydroxychloroquine may decrease the antiviral activity of remdesivir; the co-administration of these drugs is not recommended. For a limited period, the FDA had issued an Emergency Use Authorization (EUA) for hydroxychloroquine and chloroquine. This EUA was revoked in June 2020.[102]

As part of convalescent plasma therapy, plasma from recovered patients is utilized to treat patients with infections. It has also been used to treat SARS, MERS, and other viral infections. Although convalescent plasma shows promising results in many cases, all in all, the evidence to support its use in the treatment of COVID-19 is not sufficient.[54]

The neutralizing monoclonal antibodies bamlanivimab and etesevimab, administered in combination, have been shown to benefit certain subpopulations after exposure to SARS-CoV-2. Unlike vaccine-derived immunity that develops over time, the administration of neutralizing monoclonal antibodies is immediate and passive immunotherapy, with the potential to reduce disease progression, emergency room visits, hospitalizations, and death. Approvals comparable to the US EUA have been granted in several countries worldwide. Bamlanivimab and etesevimab are neutralizing immunoglobulin G1 (IgG1) monoclonal antibodies that bind to distinct yet overlapping epitopes of the SARS-CoV-2 spike protein, thereby minimizing the potential for treatment-emergent resistant

8 Treatment 55

TABLE 1 FDA-approved treatments for COVID-19.[100]

Compound (trade name in brackets)	Abbreviated treatment indication
Actemra	COVID-19 in hospitalized patients ≥ 2 years treated with systemic corticosteroids that need supplemental oxygen, mechanical ventilation, or ECMO
Sotrovimab	Mild-to-moderate COVID-19 in patients ≥ 12 years and ≥ 40 kg that tested positive for SARS-CoV-2 and at high risk for progression to severe COVID-19, including hospitalization or death
Propofol-Lipuro 1%	Sedation via continuous infusion in patients ≥ 16 years with suspected or confirmed COVID-19 who need mechanical ventilation in an ICU
Bamlanivimab and Etesevimab	Mild-to-moderate COVID-19 in patients ≥ 12 years and ≥ 40 kg at high risk for progressing to severe COVID-19 and/or hospitalization
Casirivimab and Imdevimab (REGEN-COV)	Administered together for mild to moderate COVID-19 in patients ≥ 12 years and ≥ 40 kg at high risk for progressing to severe COVID-19 and/or hospitalization
Baricitinib	COVID-19 in hospitalized patients ≥ 2 years requiring supplemental oxygen, mechanical ventilation, or ECMO
COVID-19 convalescent plasma	Hospitalized patients with COVID-19
REGIOCIT replacement solution	Patients ≥ 18 years treated with CRRT for whom regional citrate anticoagulation is appropriate in a critical care setting
Fresenius Kabi Propoven 2%	Maintain sedation via continuous infusion in patients ≥ 16 years with suspected or confirmed COVID-19 who require mechanical ventilation in an ICU setting
Remdesivir	Hospitalized COVID-19 patients ≥ 3.5 kg. For additional information, the FDA refers to the approval documentation of this remdesivir[101]
Fresenius Medical, multiFiltrate PRO System, and multiBic/multiPlus Solutions	CRRT for patients in an acute care environment during the COVID-19 pandemic

Abbreviations: *CRRT*, continuous renal replacement therapy; *ECMO*, extracorporeal membrane oxygenation; *ICU*, intensive care unit.

variants. As shown in Table 1, also further neutralizing monoclonal antibodies have been granted EUAs, showing that this class of treatment is an effective intervention for patients at risk. Apart from older age, other conditions including overweight and comorbidities such as diabetes, chronic kidney disease, chronic lung disease, and some neurological disorders, among others, increase the risk for mortality and hospitalization. For such patients, bamlanivimab and etesevimab in combination represent an acceptable benefit-to-risk ratio for patients to reduce the patient's risk to progress to severe COVID-19 and/or hospitalization.[103]

9 COVID-19, children, and "children"

"Child" has different meanings depending on the context. All minors are children administratively and legally. However, the body matures during puberty,[104] which is a relatively slow biological and physiological process. It does not occur on the night of the 18th birthday, which is an administrative and legal time limit. Administrative and legal limits are being used for several decades to define "pediatric" studies, although onset and completion of puberty have been noted to be accelerated in the last decades.[105] This confusion of the administrative/legal meaning of the word "child" as opposed to its physiological meaning and the associations that are triggered by the word "child" arose in the wake of the thalidomide disaster when the US introduced new legislation to ensure that new drugs could only be approved if their safety and effectiveness were proven. Clinical studies have emerged as a crucial element of this evidence, a principle that is now established and in force worldwide.[106]

From the 1950s on, toxicities were reported in preterm and term newborns that had been treated with new effective antibiotics. Over the following years, science established that the body of very young babies differs significantly from older people, not only in body size and the relationship between skin surface and body weight, but in particular in the enzyme structure of the liver, the filtration capacity of the kidneys, and other internal factors. From these observations, the new scientific branch of developmental pharmacology arose. But this development also bore the seeds of a new tragedy and catastrophe. From the beginning of the discussion about "pediatric drug development," the administrative and the physical meaning of the word "child" were mixed up, with the consequence that not the physical maturity of the affected persons was/is decisive for the approval of new drugs for "children," but administrative definitions that ignore the physical development level.[39,107–109] The used criteria are therefore *medically* and *scientifically* meaningless and flawed, but correct on a purely administrative and even legal level. This challenge is discussed in depth in a medical textbook.[39] Although this challenge is not the main focus of this book, it has several important implications for the way humanity has dealt with the COVID-19 pandemic in young people so far and continues to do so.[109,110]

9.1 The age of vaccination for "children"

The approval of COVID-19 vaccinations was granted for "adults" > 18 years of age for the AstraZeneca vaccine, and persons > 16 years of age for the vaccines of Moderna and BioNTech. These approvals were based on the age of the individuals who participated in the respective approval studies. But the used age limits do not correspond to natural maturation limits, they are arbitrary. It makes no sense that someone who is 15 years old and is physically bigger and stronger than mother or father should be refused the COVID-19 vaccination because s/he is still a "child." The FDA requires separate "pediatric" studies for the approval of vaccines, which have now taken place down to the age of 12 years. The FDA discusses the potential vaccination of children younger than 12 years before the completion of the ongoing pediatric studies extensively on its website. This statement correctly emphasizes that younger children need different doses. However, it does not question the need for separate efficacy studies that it demands separate approval in minors.[111]

The EMA goes even further. Years before the application for approval, it requires the respective manufacturer to commit to studies in "children" who are defined as younger than 18 years of age. For this, the manufacturer has to submit a "pediatric investigation plan" (PIP) well ahead of the planned marketing authorization application. The negotiation between the respective manufacturer and the EMA takes roughly 1 year.

Table 2 shows the EMA-demanded "pediatric" studies for remdesivir. An adolescent of 17 years is physiologically no longer a child. Medicines treat the body, not the legal status of a patient.

There is no physiological characteristic that would distinguish persons from birth to less than 18 years of age from those that are 18 years of age or older. The declaration of Helsinki explicitly disallows studies that have no meaningful question to answer.[113]

Table 3 shows the EMA PIP-demanded studies for the BioNTech vaccine. Again, it differentiates between administratively defined age groups as if these age limits would have any physiological significance. They

TABLE 2 EMA-demanded "pediatric" clinical studies for remdesivir.[112]

1. Open-label, single-arm study to evaluate the pharmacokinetics, safety, tolerability, and efficacy of remdesivir (RDV) in hospitalized children, from 32 weeks gestational age to less than 18 years of age, with confirmed COVID-19
2. Population PK modeling and simulation study to determine a pediatric dose/posology in pediatric subjects from 32 weeks gestational age to less than 18 years of age that should achieve the systemic exposures equivalent to that observed in adults
3. Extrapolation study of efficacy and safety of remdesivir from adult subjects to pediatric patients from 32 weeks gestational age (GA) to less than 18 years of age with confirmed COVID-19

TABLE 3 EMA-demanded "pediatric" clinical studies for the BioNTech vaccine.[114]

1. Double-blind dose-finding study of safety, tolerability, and immunogenicity of 2 different SARS-CoV-2 vaccine candidates (adults only) (part 1) and placebo-controlled efficacy, safety, and immunogenicity study of highly purified single-stranded, 5'-capped mRNA encoding full-length SARS-CoV-2 spike protein (BNT162b2) in adolescents from 12 years to less than 18 years of age (and adults) (part 2) for prevention of COVID-19
2. Double-blind, controlled, dose-finding safety, tolerability, and immunogenicity study of BNT162b2 in children and adolescents from 6 months to less than 12 years of age for prevention of COVID-19
3. Open, controlled, dose-finding, safety, and immunogenicity study of BNT162b2 in children from birth to less than 6 months of age for prevention of COVID-19
4. Open-label, uncontrolled, safety and immunogenicity study of BNT162b2 immunocompromised children from birth to less than 18 years of age for prevention of COVID-19

have not. Also, these studies do not make medical or scientific sense, but they are correct administratively and legally.

A bit more than a 100 years ago, most persons never had a document to identify themselves toward the authorities. A medical doctor who was asked to prescribe a medication or a vaccine assessed a young patient based on his/her physiological appearance, not based on his/her date of birth. The times have changed. Today a doctor will think three times whether he should vaccinate a young person who is celebrating his 12th birthday in a few days with the BioNTech vaccine. Medically, such doubts are wrong. They would be correct for a premature baby, but which sane doctor would even consider vaccinating a baby against COVID-19 without a compelling reason? Something can always go wrong, without any causal link to the vaccination. After vaccination, a young person may have an accident, be hit by a meteor, or become seriously ill with some illness. But if the doctor is sued, it will cost him valuable time, his nerves, and potentially high damages. This is one of the reasons why today's medicine has become defensive. Common sense can be dangerous to a doctor. Most will abide by regulations and guidelines and reject anything that needs to be thought about more in-depth. But as we see with the question of vaccinating young people, not everything that is administratively prescribed is actually OK.

9.2 Multisystem inflammatory syndrome (MIS)

During the beginning of the COVID-19 pandemic, a multisystem inflammatory syndrome (MIS) in children (MIS-C) was described.[115,116] This newly described syndrome has ascribed some similarities to the Kawasaki syndrome.[117,118] The US Centers for Disease Control and Prevention (CDC) published a Health Advisory warning,[119] in which they defined MIS-C as occurring in patients under 21 years of age with fever, laboratory evidence of inflammation, illness requiring hospitalization, multisystem (more than two organs) involvement, no alternative plausible diagnoses, and current or recent COVID-19 infection or COVID-19 exposure. The American Academy of Pediatrics (AAP) issued interim guidance, in which it listed potential complications and manifestations including Kawasaki disease-like features, toxic shock syndrome-like features, cytokine storm/macrophage activation hyper-inflammatory features, and more.[120]

The CDC health advisory triggered many PubMed-listed papers. As of August 20, 2021, the search terms "multisystem inflammatory syndrome COVID children" revealed 981 publications; the search terms "multisystem inflammatory syndrome COVID" 1099 publications; and the search term MIS-C 593 publications.[121] Compared to March 2021, until August 2021 the number of publications identified by these search terms has more than doubled.[110]

But the term MIS-C is flawed. The CDC health advisory warning[119] included four references: a CDC description of Kawasaki disease[118] plus three published papers.[115,116,122] Kawasaki disease primarily affects children that are younger than 5 years of age.[118] The Royal College of Pediatrics and Child Health (RCPCH) and Riphagen et al. guide diagnosis and treatment.[115,122] Verdoni et al. had reported an increased number of children in an Italian pediatric hospital.[116] A UK national consensus paper advised how to manage MIS-C.[123]

The first MIS-C reports had *assumed* a pediatric challenge.[115,116,122] Both the US CDC and the AAP defined MIS-C as occurring in patients younger than 21 years of age.[119,120] Furthermore, the AAP defines, in general, the pediatric population as up to 21 years of age and even older for those young patients with special needs.[124] The AAP definition makes sense for *administrative* purposes. However, bodily persons that are 15 or 20 years of age are no longer children. It is irrelevant if a patient is treated in an adult or a pediatric ward, provided s/he receives the same treatment. But a diagnosis that differentiates between patients younger and older than 21 years is simply flawed. The UK authors go even further. They do not bother to define "children."[122,123] For them, a child seems to be somebody seen by a pediatrician, which is a circular perspective. It may reflect the social standing of the RCPCH as a venerable, centuries-old institution. Nevertheless, this perspective is not scientific.

There have in the meantime also been reports of adults that experienced MIS as a result of COVID-19 infection. In adults, it is now referred to as MIS-A.[125-127] But if MIS occurs both in adults and administratively defined "children," it is flawed to differentiate between MIS-C and MIS-A. MIS-C and MIS-A disease have the same symptoms. It is a clinical syndrome, triggered by and related to a COVID-19 infection, in humans.

10 COVID-19 variants

Viruses constantly change through mutation. Virus variants have one or more mutations that differentiate them from other viruses in circulation. As expected, multiple variants of SARS-CoV-2 have emerged during the pandemic and circulated the world. All of them have slightly different proteins on their external spikes. Viral mutations and variants in the United States are routinely monitored through sequence-based surveillance, laboratory studies, and epidemiological investigations. A US government SARS-CoV-2 Interagency Group has developed a classification scheme that defines three classes of SARS-CoV-2 variants: variants of interest, variants of concern, and variants of high consequence. At present, existing variants have been classified either as variants of interest or variants of concern.[128] Variants of concern are variants associated with greater transmissibility, altered virulence, or the ability to escape natural infection- and vaccine-mediated immunity or current diagnostic tests. So far, no variant has been classified as a variant of high consequence.[128]

Also, the WHO has established a website that gives an overview of the variants.[129] Some of the new variants are reported to be more contagious and transmissible than the original SARS-CoV-2 virus type, others even as being more severe or lethal.

No binding, consistent international nomenclature has been established for SARS-CoV-2 genetic variants yet. In the beginning, the media, governments, and others referred often to new variants by the country in which they were first identified. In May 2021, the WHO announced Greek-letter names for important strains, so they could be easily referred to in a non-stigmatizing way without discriminating countries where the respective variant had first been observed.[129] However, other authors use other classification systems.[130]

Viral genomes are continuously sequenced and shared via GISAID,[45,46] facilitating the monitoring of genomic variants across the world. The SARS-CoV-2 virus is continuously evolving to escape immunity. It might result at the need for annually updated vaccines. The emergence of new variants might furthermore result in the end of the COVID-19 pandemic to come later than expected. New variants can carry the risk to be more transmissible, virulent, pathogenic, or evading immunity induced by vaccination or previous infection. However, many of these developments are unpredictable. What is certain, however, is that the longer the pandemic lasts, the higher the probability for the generation of new variants. Controlling rapidly the pandemic worldwide by a combination of a well-organized vaccination campaign coupled with strict sanitary measurements are key factors to prevent extending the pandemic is, therefore, more crucial than was visible at the beginning of the pandemic.[131]

11 Conclusions

The SARS-CoV-2 virus and the resulting COVID-19 pandemic represent a new type of healthcare challenge for mankind. Advanced science and technology have made the analysis and publication of the SARS-CoV-2 genome possible, making it immediately available to researchers, government agencies, and international institutions and organizations around the world. In a bizarre combination of lack of standing by the political leadership and at the same time determined mobilization of funds, administrative support, and encouragement to develop medicines, vaccines, and diagnostics against the SARS-CoV-2 virus, the US has technically put itself at the forefront of the fight against COVID-19, while its political—and elected—leadership has made a fool of itself at the human and scientific level. A country like the USA has such structural, material, and immaterial reserves, including intellectual and financial ones, that it can outlive even a president and his administration which were rightly be described as dangerously incompetent.[62]

In this confusing puzzle, where medieval beliefs and state-of-the-art technology sometimes fight side by side, sometimes fight each other in bitter form, it is not easy for the individual to find a compass to establish an own opinion. The mobilization of reserves in the emergencies of World War II and the COVID-19 pandemic has rightly been compared with the US government's decisions to create the Manhattan Project and to establish Operation Warp Speed. The forces behind the Manhattan Project have brought two things above all to mankind: the industrial production of penicillin, and the development of the atomic bomb. Three-quarters of a century have passed since the last atomic bomb was detonated in anger. The world has continued without a new world war. During this time the world has seen further previously unimaginable technical progress.

On a rather technical level, the development of effective vaccines against the COVID-19 pandemic has been done at a speed almost fast than light when we consider the traditional development time of 10 years and more for traditional vaccines. The term "Operation Warp Speed" was well chosen. The main challenges in dealing with the COVID-19 pandemic were less the technological limitations that we have to deal with. They exist, but they can be overcome on a technical level. The true challenges are in those areas that we would rather not see, and where corrections are necessary, but will also create resistance.

Separate efficacy studies in adolescents were scientifically not justified, but there are strong powerful associations and interest groups that benefit from such "pediatric" studies. Also, separate efficacy studies in children younger than 12 years of age are scientifically not justified. What we need for these young persons are adequate doses. If there would be serious safety concerns, registries would be adequate, but not separate efficacy studies.[39,107,109,110]

The hesitancy of the EU leadership in ordering enough vaccine shots has so far been mainly ascribed to the complex decision-making in the EU.[82,83] This criticism is correct, but there are other, deeper reasons why the US and other countries such as Israel had fewer problems with ordering enough vaccine doses than the EU. It is a fascinating mix of conflicts of interest, blind spots, arrogance, over-confidence, and over-believing in the wisdom of authorities.

In the following chapters, we analyze the crucial interfaces at which humanity is currently finding it particularly difficult to transform what is technically feasible into reality.

References

1. Worldometer. *COVID-19 Coronavirus Pandemic.* https://www.worldometers.info/coronavirus/.
2. Our World in Data. *Coronavirus (COVID-19) Deaths.* https://ourworldindata.org/covid-deaths.
3. Spitzer M. *Pandemie: Was die Krise mit uns macht und was wir daraus machen.* Munich, Germany: mvg Verlag; 2020.
4. Kekulé A. *Der Corona-Kompass: Wie wir mit der Pandemie leben und was wir daraus lernen können.* Berlin, Germany: Ullstein Verlag; 2020.
5. Perronne C. *Y a-t-il une erreur qu'ils n'ont pas commise?* Paris, France: Albin Michel Publishers; 2020.
6. Schwab K, Malleret T. *COVID-19: The Great Reset.* Geneve, Switzerland: World Economic Forum; 2020.
7. Wikipedia. *Ribosome.* https://en.wikipedia.org/wiki/Ribosome.
8. US National Library of Science. Profiles in Science. F. Crick. The Discovery of the Double Helix 1951–1953. n.d. https://profiles.nlm.nih.gov/spotlight/sc/feature/doublehelix.
9. Watson JD, Crick FH. Molecular structure of nucleic acids. A structure for deoxyribose nucleic acid. *Nature.* 1953;171(4356):737–738.
10. Watson JD, Crick FH. Genetic implications of the structure of deoxyribonucleic acid. *Nature.* 1953;171(4361):964–967.
11. Wikipedia. *History of RNA Biology.* https://en.wikipedia.org/wiki/History_of_RNA_biology.
12. Damase TR, Sukhovershin R, Boada C, et al. The limitless future of RNA therapeutics. *Front Bioeng Biotechnol.* 2021;9, 628137. https://www.frontiersin.org/articles/10.3389/fbioe.2021.628137/full.
13. Sullenger BA, Nair S. From the RNA world to the clinic. *Science.* 2016;352(6292):1417–1420. https://www.ncbi.nlm.nih.gov/pmc/articles/PMC6035743/pdf/nihms978056.pdf.
14. Kaczmarek JC, Kowalski PS, Anderson DG. Advances in the delivery of RNA therapeutics: from concept to clinical reality. *Genome Med.* 2017;9(1):60. https://www.ncbi.nlm.nih.gov/pmc/articles/PMC5485616/pdf/13073_2017_Article_450.pdf.
15. Burnett JC, Rossi JJ. RNA-based therapeutics: current progress and future prospects. *Chem Biol.* 2012;19(1):60–71. https://www.ncbi.nlm.nih.gov/pmc/articles/PMC3269031/pdf/nihms346028.pdf.

16. Kowalski PS, Rudra A, Miao L. Delivering the messenger: advances in technologies for therapeutic mRNA delivery. *Mol Ther.* 2019;27(4):710–728. https://www.ncbi.nlm.nih.gov/pmc/articles/PMC6453548/pdf/main.pdf.
17. National Institutes of Health (NIH) National Human Genome Research Institute. *Codon.* https://www.genome.gov/genetics-glossary/Codon.
18. Vasireddy D, Atluri P, Malayala S, et al. Review of COVID-19 vaccines approved in the United States of America for emergency use. *J Clin Med Res.* 2021;13(4):204–213. https://www.ncbi.nlm.nih.gov/pmc/articles/PMC8110223/pdf/jocmr-13-204.pdf.
19. Valencia DN. Brief review on COVID-19: the 2020 pandemic caused by SARS-CoV-2. *Cureus.* 2020;12(3), e7386. https://www.ncbi.nlm.nih.gov/pmc/articles/PMC7179986/pdf/cureus-0012-00000007386.pdf.
20. Goyal M, Tewatia N, Vashisht H, et al. Novel corona virus (COVID-19); global efforts and effective investigational medicines: a review. *J Infect Public Health.* 2021;14(7):910–921. https://www.ncbi.nlm.nih.gov/pmc/articles/PMC8088038/pdf/main.pdf.
21. Ganesh B, Rajakumar T, Malathi M, et al. Epidemiology and pathobiology of SARS-CoV-2 (COVID-19) in comparison with SARS, MERS: an updated overview of current knowledge and future perspectives. *Clin Epidemiol Glob Health.* 2021;10:100694. https://www.ncbi.nlm.nih.gov/pmc/articles/PMC7806455/pdf/main.pdf.
22. Wikipedia. *Li Wenliang.* https://en.wikipedia.org/wiki/Li_Wenliang.
23. Wikipedia. *COVID-19 Pandemic.* https://en.wikipedia.org/wiki/COVID-19_pandemic.
24. Wikipedia. *COVID-19 Pandemic in Mainland China.* https://en.wikipedia.org/wiki/COVID-19_pandemic_in_mainland_China.
25. García LF. Immune response, inflammation, and the clinical spectrum of COVID-19. *Front Immunol.* 2020;11:1441. https://www.ncbi.nlm.nih.gov/pmc/articles/PMC7308593/pdf/fimmu-11-01441.pdf.
26. Pascarella G, Strumia A, Piliego C, et al. COVID-19 diagnosis and management: a comprehensive review. *J Intern Med.* 2020;288(2):192–206. https://www.ncbi.nlm.nih.gov/pmc/articles/PMC7267177/pdf/JOIM-9999-na.pdf.
27. Wikipedia. *COVID-19.* https://en.wikipedia.org/wiki/COVID-19.
28. Platto S, Xue T, Carafoli E. COVID19: an announced pandemic. *Cell Death Dis.* 2020;11(9):799. https://www.ncbi.nlm.nih.gov/pmc/articles/PMC7513903/pdf/41419_2020_Article_2995.pdf.
29. Li Z, Jiang J, Ruan X, et al. The zoonotic and natural foci characteristics of SARS-CoV-2. *J Biosaf Biosecur.* 2021;3(1):51–55. https://www.ncbi.nlm.nih.gov/pmc/articles/PMC8221912/pdf/main.pdf.
30. Kadam SB, Sukhramani GS, Bishnoi P, et al. SARS-CoV-2, the pandemic coronavirus: molecular and structural insights. *J Basic Microbiol.* 2021;61(3):180–202. https://www.ncbi.nlm.nih.gov/pmc/articles/PMC8013332/pdf/JOBM-61-180.pdf.
31. Mallapaty S. What's next in the search for Covid's origins. World Health Organization report makes a reasonable start, scientists say, but many questions remain. *Nature.* 2021;592:337–338. https://media.nature.com/original/magazine-assets/d41586-021-00877-4/d41586-021-00877-4.pdf.
32. Tiezzi S. *WHO's COVID-19 Origin Report Leaves Everyone Unsatisfied. The U.S. led its allies in issuing a statement of concern, while China attempted to push back on pointed comments from the WHO's director general.* Thediplomat; April 1, 2021. https://thediplomat.com/2021/04/whos-covid-19-origin-report-leaves-everyone-unsatisfied/.
33. Hernández JC, Gorman J. *On W.H.O. Trip, China Refused to Hand Over Important Data.* The New York Time; February 12, 2021. https://www.nytimes.com/2021/02/12/world/asia/china-world-health-organization-coronavirus.html.
34. WHO. *Origins of the SARS-CoV-2 Virus. WHO-Convened Global Study of the Origins of SARS-CoV-2 (Including Annexes);* March 30, 2021. https://www.who.int/health-topics/coronavirus/origins-of-the-virus.

35. de Jesús EG. *4 Takeaways From the WHO's Report on the Origins of the Coronavirus. The Leading Hypothesis Is It Spread to People From Bats via a Yet-to-be-Identified Animal.* ScienceNews; April 1, 2021. https://www.sciencenews.org/article/covid-coronavirus-origin-who-report-takeaways.
36. Cullinan K. *China Has Failed to Share 'Raw Data' About Virus Origin so Lab Accident can't be Ruled Out, Says Tedros.* Health Policy Watch; July 15, 2021. https://healthpolicy-watch.news/china/.
37. Greenwood B. The contribution of vaccination to global health: past, present and future. *Philos Trans R Soc Lond Ser B Biol Sci*. 2014;369(1645):20130433. https://www.ncbi.nlm.nih.gov/pmc/articles/PMC4024226/pdf/rstb20130433.pdf.
38. Colón AR. *Nurturing Children. A History of Pediatrics*. Westport, CT: Greenwood Press; 1999.
39. Rose K. *Considering the Patient in Pediatric Drug Development. How Good Intentions Turned into Harm*. London: Elsevier; 2020. https://www.elsevier.com/books/considering-the-patient-in-pediatric-drug-development/rose/978-0-12-823888-2. https://www.sciencedirect.com/book/9780128238882/considering-the-patient-in-pediatric-drug-development.
40. Rappuoli R, Pizza M, Del Giudice G, et al. Vaccines, new opportunities for a new society. *Proc Natl Acad Sci U S A*. 2014;111(34):12288–12293.
41. Piot P, Larson HJ, O'Brien KL, et al. Immunization: vital progress, unfinished agenda. *Nature*. 2019;575(7781):119–129.
42. Wikipedia. *Expanded Program on Immunization*. https://en.wikipedia.org/wiki/Expanded_Program_on_Immunization.
43. GAVI. *Our Alliance*. https://www.gavi.org/our-alliance.
44. Wikipedia. *GAVI*. https://en.wikipedia.org/wiki/GAVI.
45. *Global Initiative on Sharing All Influenza Data (GISAID)*. www.gisaid.com.
46. Wikipedia. *GISAID*. https://en.wikipedia.org/wiki/GISAID.
47. Wikipedia. *Global Influenza Surveillance and Response System (GISRS)*. https://en.wikipedia.org/wiki/Global_Influenza_Surveillance_and_Response_System.
48. WHO. *Global Influenza Surveillance and Response System (GISRS)*; 2021. https://www.who.int/initiatives/global-influenza-surveillance-and-response-system.
49. WHO. *Genomic Sequencing of SARS-CoV-2: A Guide to Implementation for Maximum Impact on Public Health*; 2021. https://www.who.int/publications/i/item/9789240018440.
50. *Virological—Discussion Forum for Analysis of Virus Genomes*. https://virological.org.
51. Holmes E. *Novel 2019 Coronavirus Genome*; January 10, 2020. https://virological.org/t/novel-2019-coronavirus-genome/319.
52. Zhang EZ, Holmes EC. A genomic perspective on the originand emergence of SARS-CoV-2. *Cell*. 2020;181(2):223–227. https://www.ncbi.nlm.nih.gov/pmc/articles/PMC7194821/pdf/main.pdf.
53. Singh J, Pandit P, McArthur AG, et al. Evolutionary trajectory of SARS-CoV-2 and emerging variants. *Virol J*. 2021;18(1):166. https://www.ncbi.nlm.nih.gov/pmc/articles/PMC8361246/pdf/12985_2021_Article_1633.pdf.
54. Rehman SU, Rehman SU, Yoo HH. COVID-19 challenges and its therapeutics. *Biomed Pharmacother*. 2021;142, 112015. https://www.ncbi.nlm.nih.gov/pmc/articles/PMC8339548/pdf/main.pdf.
55. Wikipedia. *Travel Restrictions Related to the COVID-19 Pandemic*. https://en.wikipedia.org/wiki/Travel_restrictions_related_to_the_COVID-19_pandemic.
56. Chinazzi M, Davis JT, Ajelli M, et al. The effect of travel restrictions on the spread of the 2019 novel coronavirus (COVID-19) outbreak. *Science*. 2020;368(6489):395–400. https://www.ncbi.nlm.nih.gov/pmc/articles/PMC7164386/pdf/368_395.pdf.
57. Hollingsworth J. *New Zealand and Australia Were Covid Success Stories. Why Are They Behind on Vaccine Rollouts?* CNN; April 16, 2021. https://edition.cnn.com/2021/04/15/asia/new-zealand-australia-covid-vaccine-intl-dst-hnk/index.html.
58. The Lowy Institute. https://www.lowyinstitute.org/.

59. Wikipedia. *COVID-19 Pandemic in the United States*. https://en.wikipedia.org/wiki/COVID-19_pandemic_in_the_United_States.
60. Landis A. *Pandemic Presidencies: How Donal Trump and Joe Biden Navigated a Year of COVID-19*. SpectrumNews1; March 9, 2021. https://spectrumlocalnews.com/nys/buffalo/news/2021/03/09/pandemic-presidencies-how-donald-trump-and-joe-biden-navigated-a-year-of-covid.
61. Anderson M. *Eight Drugs Trump Has Been Given for His COVID-19 Treatment*; October 5, 2020. https://www.beckershospitalreview.com/pharmacy/8-drugs-trump-has-been-given-for-his-covid-19-treatment.html.
62. The Editors. Dying in a leadership vacuum. *N Engl J Med*. 2020;383:1479–1480. https://www.nejm.org/doi/pdf/10.1056/NEJMe2029812?articleTools=true.
63. Wikipedia. *New Engl J Med*. https://en.wikipedia.org/wiki/The_New_England_Journal_of_Medicine.
64. Wikipedia. *Warp Drive*. https://en.wikipedia.org/wiki/Warp_drive.
65. Wikipedia. *Operation Warp Speed*. https://en.wikipedia.org/wiki/Operation_Warp_Speed.
66. Ho RJY. Warp-speed Covid-19 vaccine development: beneficiaries of maturation in biopharmaceutical technologies and public-private partnerships. *J Pharm Sci*. 2021;110(2):615–618. https://www.ncbi.nlm.nih.gov/pmc/articles/PMC7671640/pdf/main.pdf.
67. Lancet Commission on COVID-19 Vaccines and Therapeutics Task Force Members. Operation warp speed: implications for global vaccine security. *Lancet Glob Health*. 2021;9(7):e1017–e1021. https://www.ncbi.nlm.nih.gov/pmc/articles/PMC7997645/pdf/main.pdf.
68. Hotez PJ, Nuzhath T, Callaghan T, et al. COVID-19 vaccine decisions: considering the choices and opportunities. *Microbes Infect*. 2021;23(4–5):104811. https://www.ncbi.nlm.nih.gov/pmc/articles/PMC7968147/pdf/main.pdf.
69. Slaoui M, Hepburn M. Developing safe and effective covid vaccines—operation warp speed's strategy and approach. *N Engl J Med*. 2020;383(18):1701–1703.
70. Ljungberg K, Isaguliants M. DNA vaccine development at pre- and post-operation warp speed. *Vaccines (Basel)*. 2020;8(4):737. https://www.ncbi.nlm.nih.gov/pmc/articles/PMC7761981/pdf/vaccines-08-00737.pdf.
71. Khehra N, Padda I, Jaferi U, et al. Tozinameran (BNT162b2) vaccine: the journey from pre-clinical research to clinical trials and authorization. *AAPS PharmSciTech*. 2021;22(5):172. https://www.ncbi.nlm.nih.gov/pmc/articles/PMC8184133/pdf/12249_2021_Article_2058.pdf.
72. Tregoning JS, Flight KE, Higham SL, et al. Progress of the COVID-19 vaccine effort: viruses, vaccines and variants versus efficacy, effectiveness and escape. *Nat Rev Immunol*. 2021;21:1–11. https://www.ncbi.nlm.nih.gov/pmc/articles/PMC8351583/pdf/41577_2021_Article_592.pdf.
73. Chong WC, Chellappan DK, Shukla SD, et al. An appraisal of the current scenario in vaccine research for COVID-19. *Viruses*. 2021;13(7):1397. https://www.ncbi.nlm.nih.gov/pmc/articles/PMC8310376/pdf/viruses-13-01397.pdf.
74. Sharma K, Koirala A, Nicolopoulos K, et al. Vaccines for COVID-19: where do we stand in 2021? *Paediatr Respir Rev*. 2021;39:22–31. S1526-0542(21)00065-8.
75. Nance KD, Meier JL. Modifications in an emergency: the role of N1-methylpseudouridine in COVID-19 vaccines. *ACS Cent Sci*. 2021;7(5):748–756. https://www.ncbi.nlm.nih.gov/pmc/articles/PMC8043204/pdf/oc1c00197.pdf.
76. Hilts PJ. *Protecting America's Health: The FDA, Business, and One Hundred Years of Regulation*. New York, USA: Alfred A Knopf Publishers; 2003.
77. FDA. *COVID-19 Vaccines*; 2021. https://www.fda.gov/emergency-preparedness-and-response/coronavirus-disease-2019-covid-19/covid-19-vaccines.
78. *FDA Approves First COVID-19 Vaccine. Approval Signifies Key Achievement for Public Health*. FDA News Release; August 23, 2021. https://www.fda.gov/news-events/press-announcements/fda-approves-first-covid-vaccine.

79. EMA. *COVID-19 Vaccines*; 2021. https://www.ema.europa.eu/en/human-regulatory/overview/public-health-threats/coronavirus-disease-covid-19/treatments-vaccines/covid-19-vaccines.
80. EMA. *COVID-19 Vaccines: Authorised*; 2021. https://www.ema.europa.eu/en/human-regulatory/overview/public-health-threats/coronavirus-disease-covid-19/treatments-vaccines/vaccines-covid-19/covid-19-vaccines-authorised.
81. Oakes K. *MHRA Okays Third COVID-19 Vaccine*. Regulatory Affairs Professional Society (RAPS) Regulatory Focus; January 8, 2021. https://www.raps.org/news-and-articles/news-articles/2021/1/mhra-okays-third-covid-19-vaccine.
82. Amaro S. *Europe's Stumbling Vaccine Rollout Provides a Lesson in EU Politics*. CNBC; April 2, 2021. https://www.cnbc.com/2021/04/02/europe-covid-vaccine-slow-rollout-gives-lesson-in-eu-politics.html.
83. Gros D. *The Roots of the EU's Vaccine Debacle*. Project Syndicate; April 6, 2021. https://www.project-syndicate.org/commentary/eu-covid19-vaccination-failure-by-daniel-gros-2021-04.
84. Kurup D, Schnell MJ. SARS-CoV-2 vaccines—the biggest medical research project of the 21st century. *Curr Opin Virol*. 2021;49:52–57.
85. Sahin U, Karikó K, Türeci Ö. mRNA-based therapeutics—developing a new class of drugs. *Nat Rev Drug Discov*. 2014;13(10):759–780.
86. Garde D. *The Story of mRNA: How a Once-Dismissed Idea Became a Leading Technology in the Covid Vaccine Race*. Boston Globe; November 10, 2020. https://www.statnews.com/2020/11/10/the-story-of-mrna-how-a-once-dismissed-idea-became-a-leading-technology-in-the-covid-vaccine-race/.
87. Fiore K. *Want to Know More About mRNA Before Your COVID Jab? A Primer on the History, Scope, and Safety of mRNA Vaccines and Therapeutics*. MedPage Today; December 3, 2020. https://www.medpagetoday.com/infectiousdisease/covid19/89998.
88. Collins, F. Celebrating the Gift of COVID-19 Vaccines. n.d. https://directorsblog.nih.gov/2020/12/22/celebrating-the-gift-of-covid-19-vaccines/.
89. Philippidis A. *Top 10 RNA-Based Biopharmas of 2018*. Genetic Engineering & Biotechnology News; December 17, 2018. https://www.genengnews.com/a-lists/top-10-rna-based-biopharmas-of-2018/.
90. *Top 20 Medical Start-ups*; July 2, 2021. https://www.medicalstartups.org/top/mrna/.
91. Golob JL, Lugogo N, Lauring AS, et al. SARS-CoV-2 vaccines: a triumph of science and collaboration. *JCI Insight*. 2021;6(9), e149187. https://www.ncbi.nlm.nih.gov/pmc/articles/PMC8262277/pdf/jciinsight-6-149187.pdf.
92. Vashist S.K. In Vitro Diagnostic Assays for COVID-19: Recent Advances and Emerging Trends n.d.
93. FDA. *Emergency Use Authorisation*; 2021. https://www.fda.gov/emergency-preparedness-and-response/mcm-legal-regulatory-and-policy-framework/emergency-use-authorization#infoMedDev.
94. FDA. *In Vitro Diagnostics EUAs—Molecular Diagnostic Tests for SARS-CoV-2*; 2021. https://www.fda.gov/medical-devices/coronavirus-disease-2019-covid-19-emergency-use-authorizations-medical-devices/in-vitro-diagnostics-euas-molecular-diagnostic-tests-sars-cov-2.
95. FDA. *In Vitro Diagnostics EUAs—Antigen Diagnostic Tests for SARS-CoV-2*; 2021. https://www.fda.gov/medical-devices/coronavirus-disease-2019-covid-19-emergency-use-authorizations-medical-devices/in-vitro-diagnostics-euas-antigen-diagnostic-tests-sars-cov-2.
96. FDA. *Serology and Other Adaptive Immune Response Tests for SARS-CoV-2*; 2021. https://www.fda.gov/medical-devices/coronavirus-disease-2019-covid-19-emergency-use-authorizations-medical-devices/in-vitro-diagnostics-euas-serology-and-other-adaptive-immune-response-tests-sars-cov-2.
97. FDA. *In Vitro Diagnostics EUAs—IVDs for Management of COVID-19 Patients*; 2021. https://www.fda.gov/medical-devices/coronavirus-disease-2019-covid-19-emergency-use-authorizations-medical-devices/in-vitro-diagnostics-euas-ivds-management-covid-19-patients.

References

98. European Commission. *EU Health Preparedness: A Common List of COVID-19 Rapid Antigen Tests and a Common Standardised Set of Data to be Included in COVID-19 Test Result Certificates*; 2021. https://ec.europa.eu/health/sites/default/files/preparedness_response/docs/covid-19_rat_common-list_en.pdf.
99. Batista CM, Foti L. Anti-SARS-CoV-2 and anti-cytokine storm neutralizing antibody therapies against COVID-19: update, challenges, and perspectives. *Int Immunopharmacol*. 2021;(99), 108036.
100. FDA. *Emergency Use Authorisations (EUAs), Drug and Biological Therapeutic Products*; 2021. https://www.fda.gov/emergency-preparedness-and-response/mcm-legal-regulatory-and-policy-framework/emergency-use-authorization#coviddrugs.
101. FDA. *FDA's Approval of Veklury (remdesivir) for the Treatment of COVID-19—The Science of Safety and Effectiveness*; 2020. https://www.fda.gov/drugs/drug-safety-and-availability/fdas-approval-veklury-remdesivir-treatment-covid-19-science-safety-and-effectiveness.
102. *National Institutes of Health (NIH) COVID Treatment Guidelines. Chloroquine or Hydroxychloroquine and/or Azithromycin*. NIH; 2021. https://www.covid19treatmentguidelines.nih.gov/therapies/antiviral-therapy/chloroquine-or-hydroxychloroquine-and-or-azithromycin/.
103. Nathan R, Shawa I, De La Torre I, et al. A narrative review of the clinical practicalities of Bamlanivimab and Etesevimab antibody therapies for SARS-CoV-2. *Infect Dis Ther*. 2021;10:1–15. https://www.ncbi.nlm.nih.gov/pmc/articles/PMC8353431/pdf/40121_2021_Article_515.pdf.
104. Beunen GP, Rogol AD, Malina RM. Indicators of biological maturation and secular changes in biological maturation. *Food Nutr Bull*. 2006;27(4 Suppl Growth Standard):S244–S256. https://journals.sagepub.com/doi/pdf/10.1177/15648265060274S508.
105. Komlos J, Lauderdale BE. The mysterious trend in American heights in the 20th century. *Ann Hum Biol*. 2007;34(2):206–215.
106. Rägo L, Santo B. Drug regulation: history, present and future. In: van Boxtel CJ, Santo B, Edwards IR, eds. *Drug Benefits and Risks: International Textbook of Clinical Pharmacology*. Revised 2nd ed. Wiley; 2001. https://www.who.int/medicines/technical_briefing/tbs/Drug_Regulation_History_Present_Future.pdf.
107. Rose K, Neubauer D, Grant-Kels JM. Rational use of medicine in children—the conflict of interests story. A review. *Rambam Maimonides Med J*. 2019;10(3), e0018. Review https://doi.org/10.5041/RMMJ.10371. https://www.rmmj.org.il/userimages/928/2/PublishFiles/953Article.pdf.
108. Kearns GL, Abdel-Rahman SM, Alander SW, Blowey DL, Leeder JS, Kauffman RE. Developmental pharmacology—drug disposition, action, and therapy in infants and children. *N Engl J Med*. 2003;349(12):1157–1167.
109. Rose K, Tanjinatus O, Grant-Kels JM, et al. Minors and a dawning paradigm shift in "pediatric" drug development. *J Clin Pharmacol*. 2020;61:736–739. https://accp1.onlinelibrary.wiley.com/doi/epdf/10.1002/jcph.1806.
110. Rose K, Grant-Kels JM, Ettienne E, Tanjinatus E, Striano P, Neubauer D. COVID-19 and treatment and immunization of children—the time to redefine pediatric age groups is here. *Rambam Maimonides Med J*. 2021;12:e0010. Online first https://www.rmmj.org.il/issues/online-issue/articles-online/1209.
111. FDA 2021. *FDA Will Follow the Science on COVID-19 Vaccines for Young Children*; September 10, 2021. https://www.fda.gov/news-events/press-announcements/fda-will-follow-science-covid-19-vaccines-young-children.
112. EMA. *Remdesivir PIP EMEA-002826-PIP01-20-M02*; 2021. https://www.ema.europa.eu/en/documents/pip-decision/p/0338/2021-ema-decision-9-august-2021-acceptance-modification-agreed-paediatric-investigation-plan_en.pdf.

113. World Medical Association (WMA). *Declaration of Helsinki—Ethical Principles for Medical Research Involving Human Subjects*. Adopted by the 18th WMA General Assembly, Helsinki, Finland, June 1964, and Amended by the 64th WMA General Assembly, Fortaleza, Brazil, October 2013. https://www.wma.net/policies-post/wma-declaration-of-helsinki-ethical-principles-for-medical-research-involving-human-subjects/.
114. EMA. *BNT162b2 (Comirnaty) PIP EMEA-002861-PIP02-20-M01*. https://www.ema.europa.eu/en/documents/pip-decision/p/0179/2021-ema-decision-23-april-2021-acceptance-modification-agreed-paediatric-investigation-plan-highly_en.pdf.
115. Riphagen S, Gomez X, Gonzalez-Martinez C, et al. Hyperinflammatory shock in children during COVID-19 pandemic. *Lancet*. 2020;395(10237):1607–1608. https://www.ncbi.nlm.nih.gov/pmc/articles/PMC7204765/pdf/main.pdf.
116. Verdoni L, Mazza A, Gervasoni A, et al. An outbreak of severe Kawasaki-like disease at the Italian epicentre of the SARS-CoV-2 epidemic: an observational cohort study. *Lancet*. 2020;395(10239):1771–1778. https://www.ncbi.nlm.nih.gov/pmc/articles/PMC7220177/pdf/main.pdf.
117. Jiang L, Tang K, Levin M, et al. COVID-19 and multisystem inflammatory syndrome in children and adolescents. *Lancet Infect Dis*. 2020;20(11):e276–e288. https://www.ncbi.nlm.nih.gov/pmc/articles/PMC7431129/pdf/main.pdf.
118. US Centers for Disease Control and Prevention (CDC). *Kawasaki Syndrome*; 2020. https://www.cdc.gov/kawasaki/index.html.
119. US Centers for Disease Control and Prevention (CDC) Health Alert Network (HAN) May 14, 2020: HAN00432. *Multisystem Inflammatory Syndrome in Children (MIS-C) Associated With Coronavirus Disease 2019 (COVID-19)*. https://emergency.cdc.gov/han/2020/han00432.asp.
120. American Academy of Pediatrics (AAP). *Multisystem Inflammatory Syndrome in Children (MIS-C) Interim Guidance*; 2020. https://services.aap.org/en/pages/2019-novel-coronavirus-covid-19-infections/clinical-guidance/multisystem-inflammatory-syndrome-in-children-mis-c-interim-guidance/.
121. National Institutes of Health (NIH)—National Library of Medicine—PubMed. https://pubmed.ncbi.nlm.nih.gov/.
122. Royal College of Paediatrics and Child Health (RCPCH). *Guidance: Paediatric Multisystem-Inflammatory Syndrome Temporally Associated With COVID-19*; 2020. https://www.rcpch.ac.uk/resources/guidance-paediatric-multisystem-inflammatory-syndrome-temporally-associated-covid-19-pims.
123. Harwood R, Allin B, Jones CE, et al. A national consensus management pathway for paediatric inflammatory multisystem syndrome temporally associated with COVID-19 (PIMS-TS): results of a national Delphi process. *Lancet Child Adolesc Health*. 2020. S2352-4642(20)30304-7 https://www.thelancet.com/action/showPdf?pii=S2352-4642%2820%2930304-7.
124. Hardin AP, Hackell JM. Committee on practice and ambulatory medicine. Age limit of pediatrics. *Pediatrics*. 2017;140(3):e20172151. https://pediatrics.aappublications.org/content/pediatrics/140/3/e20172151.full.pdf.
125. Morris SB, Schwartz NG, Patel P, et al. Case series of multisystem inflammatory syndrome in adults associated with SARS-CoV-2 infection—United Kingdom and United States, March-August 2020. *MMWR Morb Mortal Wkly Rep*. 2020;69(40):1450–1456.
126. Theoharides TC, Conti P. COVID-19 and multisystem inflammatory syndrome, or is it mast cell activation syndrome? *J Biol Regul Homeost Agents*. 2020;34(5):1633–1636.
127. CDC. *MIS-A website*. https://www.cdc.gov/mis-c/mis-a.html.
128. US CDC. *SARS-CoV-2 Variant Classifications and Definitions*; 2021. https://www.cdc.gov/coronavirus/2019-ncov/variants/variant-info.html.

129. WHO. *Tracking SARS-CoV-2 Variants*; 2021. https://www.who.int/en/activities/tracking-SARS-CoV-2-variants/.
130. Wikipedia. *Variants of SARS-CoV-2*. https://en.wikipedia.org/wiki/Variants_of_SARS-CoV-2.
131. Martínez-Flores D, Zepeda-Cervantes J, Cruz-Reséndiz A, et al. SARS-CoV-2 vaccines based on the spike glycoprotein and implications of new viral variants. *Front Immunol.* 2021;(12), 701501. https://www.ncbi.nlm.nih.gov/pmc/articles/PMC8311925/pdf/fimmu-12-701501.pdf.

CHAPTER

5

Russian and Chinese vaccines

1 Russian vaccines

In August 2020, Russia surprised the world by approving Sputnik V, its first domestically developed COVID-19 vaccine, before phase III clinical trials had even begun. When the World Health Organization (WHO) declared COVID-19 a pandemic in March 2020, the Gamaleya National Center of Epidemiology and Microbiology in Moscow was already working on a vaccine, using two different common cold viruses (adenoviruses rAd26 and rAd5) as vectors to be administered separately in two injections, the second to be given 3 weeks after the first one. The rationale behind this approach was that using the same adenovirus for both injections might trigger an immune response against the vector. Then, the immune system might attack and destroy the second vaccine upon administration before it could boost the immune response against the SARS-CoV-2 virus. The vector adenoviruses are combined with the SARS-CoV-2 spike protein with the aim to trigger an immune response to the SARS-CoV-2 virus in the vaccinated person. In September 2020, the results of phase I and II open, non-randomized studies with altogether 76 participants were published in the highly recognized journal Lancet.[1] Interim phase III data were published in February 2021.[2] The randomized, double-blind, placebo-controlled trial included almost 22,000 persons 18-year-old or older treated in Moscow between September and November 2020. The early approval was met with criticism in mass media and discussions in the scientific community as to whether approval was justified in the absence of robust scientific research confirming safety and efficacy.[3,4] In September 2020, an open letter co-signed by 30 scientists worldwide raised concerns about inconsistencies in the first study that required access to the original data to fully investigate.[5] The study team dismissed the concerns in an open letter.[6] December 2, 2020, the same day the UK approved the Pfizer/BioNTech vaccine, the Russian president Vladimir Putin declared that vaccination with Sputnik V would begin immediately and would be given free to citizens in Russia. But the administration had already started earlier. Gamaleya employees had received prototype doses in spring 2020 before the phase I and II trials. Over 2500 Russian soldiers have also received the injections. In the EU, only Hungary has approved the vaccine for emergency use, but worldwide, many countries have granted emergency approval. Russia first wrongly declared it had submitted Sputnik V for EMA approval, which the EMA denied.[7] A month later, the marketing authorization application was submitted to the EMA, which now confirmed the start of its review,[3,8] which is ongoing as of September 2021. The Sputnik-V interim efficacy analysis claimed efficacy of 91.8% in the age group of 60 years and older.[2] In May 2021, an international group of biostatisticians questioned the efficacy results, highlighting data discrepancies, substandard reporting, apparent errors and numerical inconsistencies, and a very unlikely homogeneity in vaccine efficacy across age groups. The authors had made multiple independent requests for access to the raw dataset, but these were never answered. Despite publicly denying some problems, formal corrections were made to the article, thus addressing some concerns.[9]

A second Russian vaccine, EpiVacCorona, is a peptide-based vaccine developed by the Russian VECTOR Center of Virology. It consists of three chemically synthesized short fragments of a viral spike protein that are bound to a large carrier protein. A clinical trial to show whether it protects people from COVID-19 or not, was launched in November 2020. Its interim results are expected for late 2021 or early 2022. There are serious concerns about the vaccine immunogenicity data. Nevertheless, EpiVacCorona has received vaccine emergency authorization in a form of government registration in Russia and is available for vaccination outside the clinical trials.[10]

Until October 2021, neither the WHO, nor the EMA, nor the FDA has approved the Sputnik V vaccine (or any other Russian vaccine). Nevertheless, almost 70 individual countries have granted emergency approval. The WHO has requested more data from the Gamaleya Institute; inspections by the WHO of Russia's vaccine-manufacturing and clinical-trial facilities are ongoing. Evidence from Russia and other countries now suggests it is safe and effective, but questions remain about the quality of surveillance for possible rare side effects. Furthermore, as long as Sputnik V is not approved by the EMA or the WHO, it cannot be distributed through the COVID-19 Vaccines Global Access (COVAX) initiative.[11] The numbers given officially by Russia should be treated with a pinch of skepticism. In September 2021, the WHO suspended the approval process for the Sputnik V COVID-19 vaccine due to "manufacturing" concerns.[12]

Several more vaccines are in development in Russia, but none has so far gone through the required clinical trials and has been approved.[4]

Despite government propaganda, average Russians do not trust Sputnik V or EpiVacCorona. Russia's vaccination rate was just 18% in

August 2021, one of the lowest in Europe and behind many countries including Azerbaijan, Cuba, El Salvador, and Morocco. To increase the vaccination rate, Russia had rolled out requirements and tried to incentivize vaccination, including giving vaccinated citizens easier access to restaurants and recreational activities. Proof of vaccination was to be uploaded to a central database. Citizens then received a specialized QR code. However, instead of increasing vaccination rates, this QR code policy led to new corruption schemes. Russians found ways around, including purchasing doctored vaccine records on the black market. Technology hiccups of the QR code further reduced the program's efficacy. The city of Moscow scrubbed the program in late July 2021.[13]

2 Chinese vaccines

China currently has five vaccines in use in its mass immunization campaign, three inactivated-virus vaccines from Sinovac and Sinopharm, a one-shot vaccine from CanSino, and one from the team of Gao Fu, director of the Chinese Center for Disease Control and Prevention (CDC), in partnership with Anhui Zhifei Longcom.[14] The two most administered vaccines are the Sinopharm vaccine Vero Cell and the Sinovac vaccine CoronaVac. China's leading vaccine makers, namely the state-owned Sinopharm and private firm Sinovac, rely on inactivated vaccine technology. These vaccines present a killed or inactivated form of COVID-19 to the body's immune system. The use of this century-old approach was probably based on the assumption that it would create fewer regulatory and production problems than newer methods.[15] The WHO has so far approved two vaccines developed by the Chinese companies Sinopharm and Sinovac for emergency use. The Sinopharm vaccine Vero Cell has an efficacy of 79% against symptomatic COVID-19 infections, but its efficacy among certain groups, such as people 60 and older, is not clear. The efficacy of the Sinovac vaccine (CoronaVac) is between around 50% to higher than 80%, depending on the country where trials were held. Vaccines based on inactivated viruses are easy to manufacture and are known for their safety, but they tend to produce a weaker immune response compared to some other vaccine types. In contrast, mRNA vaccines contain genetic instructions that tell cells to produce spike proteins that then prime the immune system.[16,17]

The efficacy of Chinese vaccines faces growing scrutiny, compounded by a lack of data on their protection against the more transmissible Delta variant that is now increasingly active worldwide. In an analysis, the news agency Consumer News and Business Channel (CNBC) found that weekly COVID-19 cases, adjusted for population, remained elevated in at least six of the world's most inoculated countries. Five of them had relied on vaccines from China, the sixth country was the UK. Many of the countries and territories that approved vaccines by Sinopharm and Sinovac are developing nations that could not get access to more effective vaccines in the initial phases of the pandemic.[16]

Vaccines made by Sinovac, a private company, and Sinopharm, a state-owned firm, comprised the majority of Chinese vaccines distributed to many countries including Mexico, Turkey, Indonesia, Hungary, Brazil, and Turkey. However, the companies did not publish peer-reviewed data on the clinical pivotal studies and have been criticized for a lack of transparency.[14] As Russia did for its Sputnik V, China skipped the final stage of clinical trials undergone by Western vaccines to roll out the Sinovac and Sinopharm vaccines faster. In Brazil, the Sinovac vaccines only halved the chance of developing symptoms of COVID (even a Chinese official commented off-hand that its efficacy is "not high"). Protection went up to 67% in Chile and 85% in Turkey.[18]

New studies have questioned the efficacy of the Chinese vaccines in comparison with other vaccines licensed by the FDA and the EMA. Several countries have begun to lessen their reliance on Chinese vaccines. Abu Dhabi, the capital of the United Arab Emirates, announced that it would require citizens fully vaccinated with Sinopharm vaccines to get a third booster dose of Sinopharm or another vaccine. Brazil and Turkey have similarly opted for booster shots. Thailand is employing mix-and-match strategies with Western jabs to bolster the efficacy of Chinese vaccines. Malaysia has begun to phase out the use of Sinovac's vaccine owing to questions about its efficacy. Studies in further countries showed that recipients of the Sinopharm vaccine, and especially older populations, faced a higher risk of breakthrough infections, hospitalizations, and deaths, compared with people vaccinated with Western vaccines. But even as Sinopharm's vaccine may not be performing as strongly as its counterparts, the data showed it to be effective in stopping many infections and deaths among those vaccinated.[19,20] As long as nothing better is available, the vaccines developed by Chinese companies are better than nothing.

In April 2021, the director of the China Centers for Disease Control and Prevention (CDC) and with that China's top disease control official admitted that the efficacy of the vaccines produced by Sinopharm and Sinovac is low and that the government is considering mixing them to give them a boost.[14]

As of September 2021, China has not administered COVID-19 vaccines that are based on mRNA technology. The mRNA technology has so far been shown to be the most effective tool in preventing the spread of COVID-19. Early in 2021, Chinese state media outlets attempted to sow doubt about mRNA vaccines. In January 2021, the *Global Times*, a nationalistic tabloid, attacked Western media for their critical reports about Chinese vaccines and hinted that relying on new mRNA vaccines might be dangerous. Days later, the *People's Daily*, the mouthpiece of the Chinese Communist Party, followed up with a story that claimed an unproven link between deaths in Norwegian nursing homes and the Pfizer/BioNTech vaccine.[15]

However, China's mRNA stance may be changing. In July 2021, Chinese media reported that Chinese regulators had completed a review of the Pfizer/BioNTech COVID-19 vaccine, distributed locally via China's Fosun Pharma. Fosun is still awaiting final approval from regulators, but, once approved, Fosun could deploy the 100 million doses it acquired from BioNTech last December to the Chinese market by the end of 2021. The approval would also unlock Fosun's capacity to produce 1 billion more BioNTech shots domestically per year, part of the deal Fosun and BioNTech struck in May to make a new joint venture company in China.[15]

The rise of the Delta variant may contribute to a strategic change. Amid Delta-driven COVID-19 outbreaks, foreign governments appear to be losing confidence in Chinese vaccines compared with mRNA vaccines. Beijing may be coming around to the idea that an mRNA vaccine could bolster its pandemic response and ease its long-awaited border reopening. China might also need to accept mRNA vaccinations to reopen its borders. Relying largely on Sinovac and Sinopharm, China has distributed over 1.4 billion vaccine doses to its citizens, enough to cover over half its population. Even with China's fast vaccinations, it may not reopen its borders until mid-2022, partially because of concerns that the Sinovac and Sinopharm vaccines may prevent deaths but be limited in their ability to prevent transmission of the virus. China may also eventually add mRNA vaccines from domestic producers. Walvax Biotechnology, a private Chinese vaccine company, has the leading mRNA vaccine candidate in China and is awaiting clearance to initiate Phase III trials.[15]

3 Assessment

Sputnik V was announced with great fanfare long before reliable data was published. When some more data came out piece by piece, many researchers worldwide criticized inconsistencies. International trust in the quality of Russian data is rightly low to nonexistent. Before the WHO or EMA allows Sputnik V, they must view the raw data and also check whether the patients specified in the raw data exist at all. The extent to which the Russian state is prepared to falsify official data was evident from the systematic doping activities of the Russian state, which led to the exclusion of Russia from the Olympic Games.[21] In Tokyo in 2021, Russian Athletes could not participate officially, but those that proved negative in the doping tests were allowed to participate under the designation "Olympic Athletes from Russia" (OAR).

Sputnik V appears to be effective,[3,11] but the certainty with which it can be proven is poor. Russia had hoped to be able to develop successful vaccination diplomacy with Sputnik V. Countries will use the Russian vaccine as long as they have no alternative. As in all areas in which Russia is involved in international affairs, great skepticism is rightly called for. This is also reflected in the great skepticism of Russian citizens toward Sputnik V.

The Chinese vaccines from Sinovac and Sinopharm are halfway effective, but not enough to permanently stop the COVID-19 pandemic. In the medium and long term, also China will probably have to use mRNA-based vaccines. China's communication first about the outbreak of the COVID-19 pandemic and then about its vaccines shows how difficult it is for a police state to cope with developments that the leadership did not expect.

There seems to be a change of position toward the company's vaccines and mRNA technology, but trust in the communication of the Chinese government and the Chinese authorities has in no way increased due to the handling of the COVID-19 pandemic. China had hoped to distinguish itself internationally as the advocate of the developing world in the fight against the COVID-19 pandemic. That seemed to work initially, in the short term, but now the world of facts is catching up with China. The efficacy of its vaccines is too low, and unlike the Chinese citizens, who are constantly exposed to the state propaganda machine and are subject to physical control by the party and the police,[22] the rest of the world can form its own opinion.

The medium and long-term way out of the COVID-19 pandemic will only be possible with highly efficient modern vaccines. Both the Chinese and Russian vaccines have not been the geopolitical wins their governments had been hoping for when they introduced them.[23]

References

1. Logunov DY, Dolzhikova IV, Zubkova OV, et al. Safety and immunogenicity of an rAd26 and rAd5 vector-based heterologous prime-boost COVID-19 vaccine in two formulations: two open, non-randomised phase 1/2 studies from Russia. *Lancet*. 2020;396:887–897. https://doi.org/10.1016/S0140-6736(20)31866-3. 32896291.
2. Logunov DY, Dolzhikova IV, Shcheblyakov DV, et al. Safety and efficacy of an rAd26 and rAd5 vector-based heterologous prime-boost COVID-19 vaccine: an interim analysis of a randomised controlled phase 3 trial in Russia. *Lancet*. 2021;397:671–681. https://www.ncbi.nlm.nih.gov/pmc/articles/PMC7852454/pdf/main.pdf.
3. Wiki. *Sputnik V COVID-19 Vaccine*. https://en.wikipedia.org/wiki/Sputnik_V_COVID-19_vaccine.
4. Baraniuk C. COVID-19: what do we know about Sputnik V and other Russian vaccines? *BMJ*. 2021;372, n743. https://www.bmj.com/content/bmj/372/bmj.n743.full.pdf.
5. Andreev K, Aubersen Y, Battino M, et al. *Note of Concern*. Cattivi Scienziati; Fighting Bad and Pseudo-Science; September 7, 2020. https://cattiviscienziati.com/2020/09/07/note-of-concern.

References

6. Logunov DY, Dolzhikova IV, Tukhvatullin AI, et al. Safety and efficacy of the Russian COVID-19 vaccine: more information needed—authors' reply. *Lancet*. 2020;396:54–55. https://www.ncbi.nlm.nih.gov/pmc/articles/PMC7503057/pdf/main.pdf.
7. EMA. *Clarification on Sputnik V Vaccine in the EU Approval Process*; February 10, 2021. www.ema.europa.eu/en/news/clarification-sputnik-v-vaccine-eu-approval-process.
8. EMA. *EMA Starts Rolling Review of the Sputnik V Covid-19 Vaccine*; March 4, 2021. www.ema.europa.eu/en/news/ema-starts-rolling-review-sputnik-v-covid-19-vaccine.
9. Bucci EM, Berkhof J, Gillibert A, et al. Data discrepancies and substandard reporting of interim data of Sputnik V phase 3 trial. *Lancet*. 2021;397(10288):1881–1883. https://www.thelancet.com/action/showPdf?pii=S0140-6736%2821%2900899-0.
10. Wikipedia. *EpiVacCorona*. https://en.wikipedia.org/wiki/EpiVacCorona.
11. Nogrady B. Mounting evidence suggests Sputnik COVID vaccine is safe and effective. *Nature*. 2021;595:339–340. https://www.nature.com/articles/d41586-021-01813-2.
12. Sputnik V. *WHO Suspends Approval Process for COVID Vaccine due to 'Manufacturing' Concerns*. Euronews; September 20, 2021. https://www.euronews.com/next/2021/09/16/sputnik-v-who-suspends-approval-process-for-covid-vaccine-due-to-manufacturing-concerns.
13. Kier G, Stronski P. *Russia's Vaccine Diplomacy Is Mostly Smoke and Mirrors. Russian Scientists Rolled Out the Country's COVID-19 Vaccine Last Summer, Beating Western Vaccine Producers to the Finish Line. But Scarce Data, Broken Promises, and Corruption Have Led the Vaccine to Lose Its Luster*. Carnegie Endowment For International Peace; August 3, 2021. https://carnegieendowment.org/2021/08/03/russia-s-vaccine-diplomacy-is-mostly-smoke-and-mirrors-pub-85074.
14. McDonald J, Wu H. *Top Chinese Official Admits Vaccines Have Low Effectiveness*. AP News; April 11, 2021. https://apnews.com/article/china-gao-fu-vaccines-offer-low-protection-coronavirus-675bcb6b5710c7329823148ffbff6ef9.
15. McGregor G. *1.4 Billion Doses Later, China Is Realizing It May Need MRNA COVID Vaccines*. Fortune; July 17, 2021. https://fortune.com/2021/07/16/china-mrna-vaccine-pfizer-biontech-fosun-doses/.
16. Lee YN. *Six Vaccinated Countries Have High Covid Infection Rates. Five of Them Rely on Chinese Vaccines*. CNBC; July 7, 2021. https://www.cnbc.com/2021/07/08/five-vaccinated-countries-with-high-covid-rates-rely-on-china-vaccines.html.
17. *China's Sinopharm Seeks to Develop Its Own mRNA Covid Vaccine. Biotech's Jab Could Give the Technology a Boost Amid Concerns Over Efficacy of Conventional Shots*. Primrose Riordan and William Langley. Financial Times; September 6, 2021. https://www.ft.com/content/91cbc7ef-808d-448a-a832-2384f045dc35.
18. *Some Nations Using China's Vaccines Are Battling Outbreaks. So Do They Work? A Quick Guide to Sinovac and Sinopharm—and What They Mean for Global Protection*. Sherryn Groch; July 7, 2013. https://www.smh.com.au/world/asia/some-nations-relying-on-china-s-vaccines-are-battling-outbreaks-so-do-the-vaccines-work-20210706-p5878h.html.
19. McGregor G. *How Do China's COVID Vaccines Fare Against the Delta Variant?* Furtone; August 31, 2021. https://fortune.com/2021/08/31/china-covid-vaccine-sinovac-sinopharm-delta-variant-effective/.
20. Pflughoeft A. *Do China's Vaccines Work? 4 Countries Tell Cautionary Tales. These Countries Relied on Chinese Vaccines. Now, They're Facing Some of the Worst Outbreaks in the World*; June 23, 2021. https://www.deseret.com/coronavirus/2021/6/23/22547032/china-coronavirus-vaccines-under-scrutiny.
21. Wikipedia. *Doping in Russia*. https://en.wikipedia.org/wiki/Doping_in_Russia.
22. Strittmatter K. We have been harmonized. In: *Life in China's Surveillance State*. New York, NY, USA: HarperCollins; 2020.
23. Bremmer I. *Why the Chinese and Russian Vaccines Haven't Been the Geopolitical Wins They Were Hoping for*; August 2, 2021. https://time.com/6086028/chinese-russian-covid-19-vaccines-geopolitics/.

CHAPTER 6

The European Union (EU) response to the COVID-19 pandemic

1 The original course of the pandemic in Europe

The first COVIC-19 case in the European Union (EU) was reported in January 2020. By May 2020, there were already more than 1 million infections and more than 100,000 deaths across the EU. COVID-19 was a completely unexpected disease for which the entire World was almost completely unprepared for including Europe. In the beginning, there was great uncertainty as to how it should go on, and major disagreements among EU member states. In the first few months, there was an overall severe shortage of protective clothing and masks, with some countries having stocked halfway good emergency supplies in the years before, other countries almost not at all.[1,2]

Caught by surprise by the spread of the SARS-CoV-2 virus across the continent, national and EU political leaders found them overtaken by an emergency situation. Pandemic diseases of such proportions had not only been expected in Europa but had been predicted and identified in the preceding weeks as an imminent threat to Europe and the rest of the world by both the World Health Organization (WHO) and the European Centre for Disease Prevention and Control.[3]

Various EU countries adopted their strategies to monitor and contain outbreaks. Italy, Spain, and Belgium introduced strict quarantines and restricted movement early in the pandemic. Sweden and the Netherlands opted for voluntary measures. While the EU had in principle long-established mechanisms to tackle outbreaks of infectious diseases, the implementation at the start of the pandemic was slow, inefficient, and hampered by bureaucracy.[4,5]

2 Details of the EU response

In hindsight, a lack of outbreak preparedness at all levels from regional, national, and EU levels was the major weakness identified in the EU. (It was not much different in most other countries, except for Israel and those countries that had the initial advantage of geographic isolation and reacted swiftly and decisively, such as Australia, New Zealand, and Taiwan.) Investment and interest in health emergency preparedness at the national level had been neglected for several years, although experts and the WHO have repeatedly warned about future pandemics. One reason was the belief that infectious diseases would break out mainly in developing countries and that the EU would be invulnerable due to its prosperity and generally stable situation.[4]

In March 2020, as the threat posed by COVID-19 became clear, the EU declared a state of emergency. But the EU does not have the legal powers to impose health management policy or actions, such as quarantine measures or closing schools, on member states.[1,2]

High-level video conferences were held in March 2020 by the European Council, a collegiate body that defines the overall political directions and priorities of the European Union. It comprises the heads of state or government of the EU member states and a few more members. The European Council had started as an informal summit and became an official institution in 2009. There was agreement on five key elements of a necessary response: limiting the spread of the virus; the need to provide medical equipment, specifically masks, and respirators; promoting research, including research into a vaccine; tackling socio-economic consequences; and helping citizens stranded in third countries.[6]

Virtually all EU states demanded some form of social distancing of the entire population in a different country-by-country regulatory mix of social distancing measures by the prohibition of public gatherings or workplace and school closures, and the introduction of travel restrictions.[3]

When US President Trump imposed a travel ban from Europe to the US, EU leaders criticized this strongly, specifically as this decision had been taken unilaterally and without consultation. But also in March 2020, the EU Commission recommended that EU citizens should remain within the EU to avoid spreading the virus in other countries. Also in March 2020, the EU closed borders to non-nationals, and most European Countries ordered a lockdown. Most national borders were closed to non-urgent traffic.[1,2]

During the COVID-19 waves in winter 2020/21, many European countries experienced high numbers of infections that in several areas overwhelmed hospitals, specifically where there was insufficient intensive care unit capacity. Even countries with relatively few cases and a low death toll during the beginnings of the pandemic were hit severely in the winter. In spring 2021, Europe went through another surge in cases that peaked in April. The emergence and severity of these waves varied greatly across Europe.

Messenger RNA (mRNA) vaccines reduce infections typically in the 80%–90% range after two doses. The administration of two different vaccines to the same person is allowed and practiced. Furthermore, vaccines reduce transmissibility even if breakthrough infections occur, i.e., an infection in a person that has been vaccinated twice. Importantly, mRNA vaccines prevent severe symptoms and hospitalization, resulting in risk reductions. With the submission of the manuscript for this book, vaccination continues in Europe. Repeatedly changing policy recommendations and constant media coverage unsettled people, especially after evidence of possible links to rare adverse, sometimes fatal, side effects emerged for the AstraZeneca and the Johnson & Johnson vaccines. Among more senior and vulnerable citizens, vaccine uptake was generally high, while in younger age groups, willingness to get vaccinated was lower. There were also concerns among health care workers about getting vaccinated. The pandemic policies were overall not well received in many European countries. Among the reasons responsible for this were continued high economic] and psychological burdens, inadequate risk communication, a lack of transparent long-term strategies from governments and the EU leadership, and increasing vaccination coverage, and a general erosion of trust.[7]

3 Future EU plans

In her State of the Union address 2020 the president of the European Commission, Ursula von der Leyen, had announced the creation of a "European Health Union" to help strengthen the EU's health security framework, and to reinforce the crisis preparedness and response role of key EU agencies.[8,9] In this address, she explained the aim to protect the health of all European citizens, that the COVID-19 pandemic had highlighted the need for more coordination in the EU, more resilient health systems, and better preparation for future crises. Therefore, she outlined, the EU is changing its way to address cross-border health threats and starts building a European Health Union to protect citizens with high-quality care in a crisis and to prevent and manage future health emergencies that affect the whole of Europe. Her announcement included revamping the existing legal framework for serious cross-border threats to health and to reinforcing the crisis preparedness and response role specifically of the two key EU agencies European Centre for Disease Prevention and Control (ECDC) and the European Medicines Agency (EMA).[9]

The ECDC is an agency of the EU with the mission to strengthen Europe's defenses against infectious diseases. It covers a wide spectrum of activities, including surveillance, epidemic intelligence, response, scientific advice, microbiology, preparedness, public health training, international relations, health communication, and the scientific journal Eurosurveillance. The ECDC was established in 2004 in reaction to the SARS outbreak in 2003 and is headquartered in Solna, Sweden.[10,11] The EMA was set up in 1995 to harmonize (but not replace) the work of the existing national medicine regulatory authorities. I work as an umbrella organization with and above the respective regulatory authorities of the EU national member states.[12]

In February 2021, the EU Commission announced the creation of the "European Health Emergency Preparedness and Response Authority (HERA) incubator," as with the COVID-19 pandemic, new challenges and threats continued to emerge, ranging from variants to vaccine adaptation or mass production. The HERA incubator was announced as the vanguard for the future HERA itself. HERA is planned to work in tight collaboration with the WHO. A primary objective of HERA incubator is to ensure for the EU swift access to the volume of vaccines needed to face virus variant threats. The HERA incubator should also initially facilitate and encourage several concurrent projects to identify and develop the most promising vaccine candidates. It should then ensure the availability of manufacturing capacity to allow production and supply at the scale of new or adapted vaccines. HERA incubator is announced to be of benefit far beyond the EU's borders. The HERA incubator activity will reach out to and cooperate with the EU external and global partners such as CEPI,[13,14] GAVI,[15,16] and the WHO[17] on the challenge of new vCOVID-19 virus variants (see also Chapter 7). In the medium- and long- term, the EU plans to cooperate with low- and middle-income countries, in particular in Africa, to help scale up local manufacturing and production capacities. The European Commission emphasized that, given the race against time, sufficient funding will be available. The HERA Incubator will start rolling out its activities immediately.[18]

In future emergencies, HERA should ensure the development, production, and distribution of medicines, vaccines, and other medical countermeasures, such as gloves, masks, and other protective clothing, that were often lacking during the first phase of the EU corona-virus response. HERA is planned to become a key pillar of the European Health Union. It thus should fill a gap in the EU health emergency response and preparedness. HERA is planned to be fully operational in early 2022. After annual reviews and adaptations, a full review is announced for 2025.[19–22]

4 A preliminary assessment of the EU response to the COVID-19 pandemic

You have to read between the lines to understand why the "European Health Union" was established in 2020, the HERA incubator in February 2021, and HERA is planned to be operational in 2022. As described in the introduction, the EU reacted slowly and sluggishly to the COVID-19 pandemic and failed to establish a leadership position worldwide, even when the US political leadership was largely paralyzed by an incompetent president.[23] It is part of the European tradition that when political structures and institutions fail, one or more new authorities are established, in this case, HERA. In the EU Commission press release, HERA will prevent, detect, and rapidly respond to health emergencies. HERA will anticipate threats and potential health crises, through intelligence gathering and building the necessary response capacities. Of course, all pronouncements emphasize that all institutions are working to the best of their ability and with full dedication. Between the lines, we read that the EU Commission knows that this was not the case. To put it provocatively: all EU institutions have failed miserably. There was a lack of protective clothing, masks, and ventilators, and when the first vaccines were finally approved in the UK and the US, the EMA had to be kicked in the rear by health ministers of EU national countries until they settled down to approve the Pfizer/BioNTech vaccine in the EU at the end of 2020.

What complicates things at the EU level is that its leaders are not directly elected. They are determined by negotiations between the governments of the EU states and are confirmed or rejected as a whole package by the EU Parliament. As a rule, retired politicians are raised to the EU level for a while, where they can give Sunday speeches without having to fear being voted out of office immediately.

It is also worth reading the description of the pandemic development from the perspective of the WHO, where we see a series of successes and wise leadership by the WHO (see also Chapters 7 and 8).[24] From the point of view of bureaucrats, things can look very different than from the point of view of the average person or the critical observer.

References

1. Wikipedia. *European Union Response to the COVID-19 Pandemic.* https://en.wikipedia.org/wiki/European_Union_response_to_the_COVID-19_pandemic.
2. Wikipedia. *COVID-19 Pandemic in Europe.* https://en.wikipedia.org/wiki/COVID-19_pandemic_in_Europe.

3. Alemanno A. *The European Response to COVID-19: From Regulatory Emulation to Regulatory Coordination?* Cambridge University Press; April 28, 2020. file:///C:/Users/klaus/AppData/Local/Temp/the-european-response-to-covid-19-from-regulatory-emulation-to-regulatory-coordination.pdf.
4. Gontariuk M, Krafft T, Rehbock C, et al. The European Union and public health emergencies: expert opinions on the Management of the First Wave of the COVID-19 pandemic and suggestions for future emergencies. *Front Public Health*. 2021;9:698995. https://www.ncbi.nlm.nih.gov/pmc/articles/PMC8417533/pdf/fpubh-09-698995.pdf.
5. Kingsland J. *EU Slow and Inefficient in Early COVID-19 Response, Says Report*. Medical News Today; August 20, 2021. https://www.medicalnewstoday.com/articles/eu-slow-inefficient-and-hampered-by-bureaucracy-in-early-covid-19-response.
6. Wikipedia. *European Council*. https://en.wikipedia.org/wiki/European_Council.
7. Iftekhar EN, Priesemann V, Balling R, et al. A look into the future of the COVID-19 pandemic in Europe: an expert consultation. *Lancet Reg Health Eur*. 2021;8:100185. https://www.ncbi.nlm.nih.gov/pmc/articles/PMC8321710/pdf/main.pdf.
8. *State of the Union Address by President von der Leyen at the European Parliament Plenary*. September 16, 2020. https://ec.europa.eu/commission/presscorner/detail/ov/SPEECH_20_1655.
9. *The Creation of a European Health Union*. Health Europa; November 12, 2020. https://www.healtheuropa.eu/the-creation-of-a-european-health-union/103859/.
10. European Centre for Disease Prevention and Control (ECDC). *An Agency of the European Union*. https://www.ecdc.europa.eu/en.
11. Wikipedia. *ECDC*. https://en.wikipedia.org/wiki/European_Centre_for_Disease_Prevention_and_Control.
12. Wikipedia. *European Medicines Agency*. https://en.wikipedia.org/wiki/European_Medicines_Agency.
13. *Coalition for Epidemic Preparedness Innovations (CEPI)*. https://cepi.net/.
14. Wikipedia. *Coalition for Epidemic Preparedness Innovations (CEPI)*. https://en.wikipedia.org/wiki/Coalition_for_Epidemic_Preparedness_Innovations.
15. Wikipedia. *GAVI*. https://en.wikipedia.org/wiki/GAVI.
16. *GAVI*. https://www.gavi.org/.
17. WHO. www.who.int.
18. EU Commission. *Communication From the Commission to the European Parliament, the European Council and the Council. HERA Incubator: Anticipating Together the Threat of COVID-19 Variants*; February 17, 2021. https://ec.europa.eu/info/sites/default/files/communication-hera-incubator-anticipating-threat-covid-19-variants_en.pdf.
19. *European Health Emergency Preparedness and Response Authority (HERA): Getting Ready for Future Health Emergencies*. EU Commission Press; September 17, 2021. https://www.pubaffairsbruxelles.eu/european-health-emergency-preparedness-and-response-authority-hera-getting-ready-for-future-health-emergencies-eu-commission-press/.
20. *European Health Emergency Preparedness and Response Authority (HERA)*. https://ec.europa.eu/info/law/better-regulation/have-your-say/initiatives/12870-European-Health-Emergency-Preparedness-and-Response-Authority-HERA-_en.
21. *European Health Emergency Preparedness and Response Authority (HERA)*. https://de.wikipedia.org/wiki/European_Health_Emergency_Response_Authority.
22. HERA. www.HERAresearchEU.eu.
23. The Editors. Dying in a leadership vacuum. *N Engl J Med*. 2020;383:1479–1480. https://www.nejm.org/doi/pdf/10.1056/NEJMe2029812?articleTools=true.
24. *One Year of WHO/Europe's Response to the COVID-19 Pandemic*. WHO; January 24, 2021. https://www.euro.who.int/en/health-topics/health-emergencies/coronavirus-covid-19/multimedia/one-year-of-whoeuropes-response-to-the-covid-19-pandemic.

CHAPTER 7

COVID-19 vaccines global access (COVAX) and more

1 Introduction

Overcoming the COVID-19 pandemic is at the same time a local challenge everywhere, and a global challenge in which the international structures of our world were and are involved. Humanity, social responsibility, and many noble aims are continuously proudly proclaimed. But there is no world government, and there are no binding international values. There are many interests that are diametrically opposed to one another, some of which are formulated openly, others rather elegantly whitewashed, and still others not addressed officially at all. COVID-19 Vaccines Global Access (COVAX) aims at giving low-to-middle-income countries access to COVID-19 vaccines. Within the international maze, there are numerous organizations including ATC-C, the group of 20 (G20), WHO, GAVI, CEPI, and more. In the assessment of COVAX, we have to distinguish political declamations from reality. COVAX has helped and is helping to a limited degree, but less than its protagonists promised and hoped initially.

2 Key international organizations

In 2020, the WHO, the European Commission, the Bill & Melinda Gates Foundation, and the French government launched the "Access to COVID-19 Tools Accelerator" or the "Global Collaboration to Accelerate the Development, Production and Equitable Access to New COVID-19 diagnostics, therapeutics and vaccines," in the following abbreviated as "ACT-C,"[1–3] in response to a call from the group of 20 (G20), an intergovernmental forum comprising 19 countries and the EU, which works to address major issues of the global economy, climate change, and sustainable development, composed of most of the world's largest economies. Together, the economies of the G20 represent about 90% of gross world product, three-quarters of international trade, two-thirds of the world's population, and half the world's land surface.[4] The ACT-C is not a new organization or a decision-making body, but more a framework for collaboration. The ACT-C comprises the four pillars vaccines, diagnostics, therapeutics, and the health systems connector. Each of these pillars is managed by several collaborating partners.[1-3] The vaccines pillar of the ACT-C was named "COVID-19 Vaccines Global Access" (COVAX).[5,6] COVAX aimed at coordinating resources to enable low-to-middle-income countries equitable access to COVID-19 vaccines. The collaborating partners in the COVAX are GAVI,[7,8] the Coalition for Epidemic Preparedness Innovations (CEPI),[9,10] and the WHO.[11]

GAVI (officially Gavi, the Vaccine Alliance) is a public-private global health partnership to increase access to immunization in poor countries. It facilitates vaccinations in developing countries by working with donor governments, the WHO, UNICEF, the World Bank, the vaccine industry in industrialized and developing countries, research and technical agencies, civil society, the Bill & Melinda Gates Foundation, and other private philanthropists. GAVI has observer status at the World Health Assembly. It was created in 2000 as a successor to the Children's Vaccine Initiative, which had been launched in 1990.[7,8]

The concept for CEPI (Coalition for Epidemic Preparedness Innovations) goes back to a paper in The New England Journal of Medicine (NEJM) in 2015 that proposed to establish a global vaccine development fund.[12] The authors described how for the Ebola virus in Africa there was a lot of basic research, but very limited incentives for the few large vaccine manufacturers to develop vaccines against relatively rare virus diseases, given the long development time and the development costs per vaccine of between half a billion and more than a billion dollars. The authors proposed a global vaccine development fund. The idea was picked up by several decision-makers and institutions, and CEPI was formally launched in 2017 at the World Economic Forum (WEF). Headquartered in Oslo, Norway, CEPI was co-founded and co-funded by the Bill and Melinda Gates Foundation, the Wellcome Trust, and several national governments. The founding mission of CEPI was equitable access in pandemics, i.e., selling vaccines to developing nations at affordable prices. CEPI works with international health authorities and vaccine developers to create vaccines for preventing epidemics. For the COVID-19 pandemic, CEPI has organized a US$ 2 billion funds in a global partnership between public, private, philanthropic, and civil society organizations for accelerated research and clinical testing of nine COVID-19 vaccine candidates, with the 2020–21 goal of supporting several candidate vaccines for full development to licensing. CEPI takes donations from public, private, philanthropic, and civil society organizations, to finance independent research projects to develop vaccines against emerging infectious diseases.[9,10]

3 COVID-19 vaccines global access (COVAX)

As part of its America First policy, the Trump administration refused to join COVAX because of its association with the WHO, from which it had declared its withdrawal in 2020. In 2021, Trump's successor Joe Biden announced that the US would remain in the WHO and would join COVAX. In February 2021, the US pledged $US 4 billion, making it the single largest contributor to the fund. The other major contributors are Germany, Japan, the UK, Canada, the EU, and several further European countries. As of June 2021, the WHO has approved the Oxford/AstraZeneca, Pfizer/BioNTech, Moderna, Sinopharm, Sinovac, and Janssen vaccines for emergency use. These vaccines can be distributed as part of COVAX.[5,6]

COVAX began distributing vaccines in February 2021. The first 100 million doses, originally promised for March, had been distributed by July 2021; by August, it had delivered 200 million vaccine doses (but it had initially aimed at 600 million). The continued shortage of COVID-19 vaccines delivered through COVAX is blamed on "vaccine nationalism" by richer nations, and on the diversion of 400 million Oxford/AstraZeneca vaccine doses, produced under license by the Serum Institute of India, for domestic use in India.[5,6]

By December 2020, more than 10 billion vaccine doses had been preordered by developed countries. The manufacturers of three vaccines then closest to global distribution—Pfizer, Moderna, and AstraZeneca—had predicted a manufacturing capacity of 5.3 billion doses in 2021, which could be used to vaccinate about 3 billion people, as the vaccines require two doses for a protective effect against COVID-19.

4 The discussion about booster shots

The standard regimen for mRNA COVID-19 vaccines is two doses, but some countries, including the US and Israel, have started administering third "booster" shots. A recent study showed that in Israelis over the age of 60 who had received a third "booster" jab, participants were much less likely to have severe COVID-19 than people in the same age group who had received only two jabs.[13] Nearly 1 million boosters have already been administered in the US since the FDA authorized the third shots of Pfizer's or Moderna's vaccines for people with weakened immune systems.[14,15] As usual, there are scientific concerns about a third vaccination.[16]

The WHO Director-General Tedros Adhanom Ghebreyesus called repeatedly for a global moratorium on booster doses to prioritize vaccinating the most at-risk people around the world who are yet to receive their first dose. But there was been little change in the global situation since then.[14,17]

5 Preliminary COVAX assessment

In principle, a fair vaccine allocation mechanism for COVAX would involve distributing doses to those who need it most. However, the political realities do not allow for such an approach. It might be smarter for WHO to include these realities in its strategy and communications.[18] We observe several different types of communication throughout the COVID-19 pandemic. Official propaganda everywhere speaks of the noble governments of the developed countries who are doing everything in their power to help their people. Well, indeed, governments have the first duty to look after the well-being of their people. If you do not, you lose the next election. Another level of communication is that the politicians of the developed countries must also include reality and communicate about it. The third level of communication is that the fact that countries must think of their citizens is now called "vaccination nationalism." This characterization is wrong. Booster shots prevent severe COVID-19 disease courses. Countries that have the opportunity to do so should also do so.

COVAX might have contributed to accelerating the development of COVID-19 vaccines,[19] together with other factors such as the US Operation Warp speed. However, COVAX was not as successful as a procurement tool for providing effective vaccines for low-income countries and low- and medium-income countries as the promises that have been made when COVAX was established. Some report critically that rich nations have given money to COVAX and paid lip service to the idea of vaccines for all while scrambling to buy up all the doses they could.[20,21] The situation is not that easy. Countries did not know which vaccine would work and which would not. Nobody knew that. Predicting the future is still a difficult business. Therefore, all vaccines that had a chance of obtaining approval were ordered beforehand. But not all of these vaccines have overcome this hurdle. Now the countries that now have a surplus of vaccines are passing it on to other countries. There is nothing wrong with such an approach.

References

1. *Access to COVID-19 Tools (ACT) Accelerator.* https://www.who.int/initiatives/act-accelerator.
2. *Access to COVID-19 Tools (ACT) Accelerator.* https://www.theglobalfund.org/en/act-accelerator/.
3. *Access to COVID-19 Tools (ACT) Accelerator.* https://en.wikipedia.org/wiki/Access_to_COVID-19_Tools_Accelerator.
4. Wikipedia, *Group of 20 (G20).* https://en.wikipedia.org/wiki/G20.
5. Wikipedia, *COVAX.* https://en.wikipedia.org/wiki/COVAX.
6. COVAX. *Working for Global Equitable Access to COVID-19 Vaccines.* https://www.who.int/initiatives/act-accelerator/covax.

7. Wikipedia, *GAVI*. https://en.wikipedia.org/wiki/GAVI.
8. GAVI. https://www.gavi.org/.
9. *Coalition for Epidemic Preparedness Innovations (CEPI)*. https://cepi.net/.
10. Wikipedia, *Coalition for Epidemic Preparedness Innovations (CEPI)*. https://en.wikipedia.org/wiki/Coalition_for_Epidemic_Preparedness_Innovations.
11. WHO. www.who.int.
12. Plotkin AS, Mahmoud AAF, Farrar J. Establishing a global vaccine-development fund. *N Engl J Med*. 2015;373(4):297–300. https://www.nejm.org/doi/pdf/10.1056/NEJMp1506820?articleTools=true.
13. Bar-On YM, Goldberg Y, Mandel M, et al. Protection of BNT162b2 vaccine booster against Covid-19 in Israel. *N Engl J Med*. 2021;385(15):1393–1400. https://www.nejm.org/doi/pdf/10.1056/NEJMoa2114255?articleTools=true.
14. FDA 12AUG2021. *Coronavirus (COVID-19) Update: FDA Authorizes Additional Vaccine Dose for Certain Immunocompromised Individuals*. Other fully vaccinated individuals do not need an additional vaccine dose right now. https://www.fda.gov/news-events/press-announcements/coronavirus-covid-19-update-fda-authorizes-additional-vaccine-dose-certain-immunocompromised.
15. Mendez R. *WHO Presses World Leaders to Hold Off on Covid Vaccine Booster Shots Through September*. CNBC; September 1, 2021. https://www.cnbc.com/2021/09/01/who-presses-world-leaders-to-hold-off-on-covid-booster-shots-through-september.html.
16. Krause PR, Fleming TR, Peto R, et al. Considerations in boosting COVID-19 vaccine immune responses. *Lancet*. 2021;398(10308):1377–1380. S0140-6736(21)02046-8 https://www.thelancet.com/action/showPdf?pii=S0140-6736%2821%2902046-8.
17. WHO. *WHO Director-General's Opening Remarks at the Media Briefing on COVID-19*; September 8, 2021. https://www.who.int/director-general/speeches/detail/who-director-general-s-opening-remarks-at-the-media-briefing-on-covid-19- - -8-september-2021.
18. Sharma S, Kawa N, Gomber A. WHO's allocation framework for COVAX: is it fair? *J Med Ethics*. 2021. medethics-2020-107152 https://www.ncbi.nlm.nih.gov/pmc/articles/PMC8042584/pdf/medethics-2020-107152.pdf.
19. Eccleston-Turner M, Upton H. International collaboration to ensure equitable access to vaccines for COVID-19: the ACT-accelerator and the COVAX facility. *Milbank Q*. 2021;99(2):426–449. https://www.ncbi.nlm.nih.gov/pmc/articles/PMC8014072/pdf/MILQ-99-426.pdf.
20. So AD, Woo J. Reserving coronavirus disease 2019 vaccines for global access: cross sectional analysis. *BMJ*. 2020;371, m4750. https://www.ncbi.nlm.nih.gov/pmc/articles/PMC7735431/pdf/bmj.m4750.pdf.
21. The Lancet. Access to COVID-19 vaccines: looking beyond COVAX. *Lancet*. 2021;397(10278):941. https://www.ncbi.nlm.nih.gov/pmc/articles/PMC7952094/pdf/main.pdf.

CHAPTER 8

International healthcare structures and COVID-19

1 WHO basics and "Public Health Emergencies of International Concern" (PHEICs)

In the 19th century, markedly increased trade and travel had, as a side effect, outbreaks of cholera and other epidemic diseases. Several International Sanitary Conferences took place between 1851 and 1938, focusing initially almost solely on cholera. Later conferences widened the areas of diseases and included yellow fever, brucellosis, leprosy, tuberculosis, and typhoid. In 1907, the International Office of Public Hygiene (Office international d'hygiène publique [OIHP]) was established to coordinate the quarantining of ships and ports to prevent the spread of plague and cholera. It was dissolved in 1946 and incorporated into the WHO when it was established as an agency of the United Nations (UN) in 1948. The WHO's immediate official predecessor which had been the League of Nations Health Organization, had worked through the interwar decades.[1,2]

Also, the first attempts to classify diseases and causes of death go back to the 19th century. The first International List of Causes of Death was adopted by the International Statistical Institute in 1893.[3] It distinguished general diseases from organ-specific diseases. Today, the international classification of diseases (ICD), maintained by the WHO, is the most widely used classification system.[4] It plays a key role in healthcare administration, insurance, drug development, and other areas of healthcare. In 1948, following World War II, the United Nations (UN) absorbed all the other international healthcare organizations and established the WHO as one of its agencies.[5,6] Tools like the ICD show how healthcare administration and operational aspects of healthcare are becoming increasingly complex for more than a hundred years. Our world is becoming more and more global.

WHO's initial priorities were malaria, tuberculosis, venereal diseases, maternal and child health, sanitary engineering, and nutrition. The organization had an initial budget of US $ 5 million. It was also involved in wide-ranging disease prevention and control efforts including mass campaigns against yaws, endemic syphilis, leprosy, and trachoma.[7] From the start, it worked with member countries to identify and address public health issues, support health research, and issue guidelines. In addition to governments, WHO coordinates with other UN agencies, donors, non-governmental organizations (NGOs), and the pharmaceutical industry. Investigating and managing disease outbreaks remained the responsibility of each country. WHO had and has no authority to police what member countries are doing.[8]

The WHO is as a UN agency officially responsible for international public health. It is an umbrella organization above the public healthcare in the respective countries. But there is no world government. The WHO represents the minimum common multiple of what the majority of the world's public health administrations and governments can agree upon. Its decision-making body is the World Health Assembly (WHA), organized once per year with an agenda prepared by the Executive Board. The WHA determines the WHO's policies, appoints the Director-General, elects and advises the WHO executive board, supervises financial policies, and reviews and approves the proposed budget.[9] In 1975, it launched the Special Program for Research and Training in Tropical Diseases (TDR).[6,10]

2 International Health regulations (IHR) and PHEICs

In 1969, the WHA adopted the "International Health Regulations" (IHR), last revised in 2005. The IHR expects governments to report upcoming contagious diseases. The IHR is a legally binding part of international law to enable a public health response to the international spread of diseases. They are the only international treaty that empowers the WHO to act as a global surveillance system. They also authorize the WHO to determine events as a "Public Health Emergency of International Concern" (PHEIC), in situations that are (1) serious, sudden, unusual, or unexpected; (2) with implications for public health beyond the affected states' borders; (3) may require immediate international action.[11,12] Table 1 lists the PHEICs declared by the WHO Director-General since 2007.

TABLE 1 PHEICs declared since 2007[12].

- 2009 H1N1 swine flu pandemic
- 2014 setbacks in global polio eradication efforts
- 2013–2016 Western African Ebola virus epidemic
- 2016 Zika virus outbreak
- 2018–19 Kivu Ebola epidemic
- 2019–21 COVID-19 pandemic

None of the diseases listed in Table 1 came even close to reaching the dimensions of the COVID-19 pandemic. Criticism of the PHEIC declarations (and non-declaration) had suggested that the rationale used was contradictory, non-transparent, or in direct violation of the IHR criteria. A first, comprehensive review revealed considerable inconsistency, and more consistency and transparency were recommended.[13]

3 The WHO's life of its own

The WHO is the highest international institution on health issues. However, executive power remains with the states. The WHO reflects the current political structure of our planet and the level of international communication. The various governments talk to each other, have diplomatic relations, and address and handle an increasing amount of international operational matters. The media play a key role in the ever-growing theater of international discussion. More and more key representatives enter the increasingly short-lived international stage. The World Economic Forum was founded in 1971 and is "committed to improving the state of the world by engaging business, political, academic, and other leaders of society to shape global, regional, and industry agendas". It hosts an annual meeting in Switzerland, brings together some 3000 business leaders, political leaders, economists, celebrities, and journalists.[14,15] There are numerous non-government organizations (NGOs) on various aspects of modern life, including doctors without borders,[16] and organizations that rescue refugees in the Mediterranean from alleged distress at sea after they have embarked on unseaworthy boats to reach Europe. The refugees alert the rescue services via mobile phone when their boats start to sink. The countries they left are not considered safe by the brave rescuers who fish them out of the sea, ship them to South European countries, and demand they be accepted.[17-19] Some countries still celebrate these "saviors" as heroes. But such crowds have arrived in Europe that the mood has turned, and many countries bordering the Mediterranean, including Greece, Malta, Italy, Spain, and France, use individual solutions to stop unwanted immigration.[20,21] There are countless other NGOs for the welfare of people, animals, the environment, and more. The supply of conflicting information is almost limitless. Most NGOs are fixated on their respective goals and think less about long-term effects, leading to unsolvable dilemmas. Today's asylum seekers have survived childhood illnesses, infectious diseases, and many other medical challenges that killed people in former times. But the structures in which they grow up are unable to offer a decent life to all of them. Furthermore, the ideas, political and legal frameworks, and the established doctrines in the developed countries correspond no longer sufficient to reality, opening opportunities for populists who long for the chance to make things worse under the guise of radical solutions or to resort to worn-out methods that have led to fundamental catastrophes in the past.

The decisive method to establish international decisions has remained war. Russia occupies and controls several territories of other countries, including Georgia, Ukraine, and Moldova.[22,23] Of course, Russia has elegant legal justifications for each of these. In 2014, Malaysia Airlines Flight 17 from the Netherlands to Malaysia was shot down over Ukraine. All 283 passengers and 15 crew were killed. Pro-Russian rebel militia claimed to have shot down a Ukrainian military airplane. When it became apparent that it had been a civilian plane, the separatists withdrew their claim and denied shooting down any aircraft. The weapon used had been a Russian Buk surface-to-air missile provided by the Russian 53rd Anti-Aircraft Missile Brigade. The names of the responsible persons are known. In 2019, the Dutch Public Prosecution Service charged four men with murder. International arrest warrants were issued. Russia denies any responsibility.[24,25] As long as the accused men remain in Russia, they are safe, unless the relatives of the murdered passengers finance a commando to remove them from Russia like the Israeli Mossad removed the Nazi criminal Adolf Eichmann from Argentina.[26] The western world, the structure of which was significantly shaped by the end of World War II, finds it difficult to distinguish appearance and propaganda from reality.

The basic dilemma of the WHO is that it does not represent a state administration with executive authority, but is limited to pronouncements, recommendations, the creation of documents, and coordinating activities as far as most member states agree and support action. The WHO was established in 1948 as a UN agency when the cold war emerged as a period of geopolitical tension between the US and the Soviet Union and their respective allies. In 1991, the Soviet Union dissolved, with Russia as its successor. It was called the "cold" war because there was no open war between the two superpowers.[27] Both supported their allies in proxy wars. Eventually, the European Union (EU) emerged in Europe. Furthermore, China became a new major power.

The WHO has achieved spectacular successes in areas where agreement could be reached, in particular the worldwide eradication of smallpox, for which vaccination is no longer required since 1980. The WHO has promoted mass campaigns against yaws, endemic syphilis, leprosy, trachoma, and other diseases. It launched a program on research, development, and training in fertility regulation and birth-control methods (leading also to a sharp controversy with the Catholic Church over the use of condoms). It initiated a program to vaccinate children worldwide against diphtheria, pertussis, tetanus, measles, poliomyelitis, and tuberculosis. It initiated campaigns for polio eradication, against maternal morbidity, and lifestyle diseases including diabetes and tobacco-related diseases.[7] But the WHO declarations are flowery, politically correct, sound nice, and represent more the self-portrayals of governments, scientific and medical institutions, and health authorities. Their connection to reality is often weak. The Declaration of Alma Ata of 1978 defines health as a fundamental human right and as "a state of complete physical, mental and social wellbeing, and not merely the absence of disease or infirmity".[28] Only a few billionaires would be healthy by this definition.

At the technical level, mankind has progressed remarkably since the cold war. We have medicines today that people could only dream of during World War II. Modern drugs have helped the developing world to reduce infectious diseases. But many factors that make up developed countries are still missing in the developing world. Public health is weak, corruption is pervasive, and there is a gigantic black market for counterfeit drugs without effective government control of drugs.

The consequences of this progress are bizarre. Child mortality has declined worldwide. Many infectious diseases that used to kill in the past are today partially prevented, and many can be treated by modern antibiotics. There is modern communication, many doctors and scientists have been trained in developed countries. Increasingly, also poor people have internet access and access to mobile phones. The political structures are difficult to grasp. The majority of this world's countries are not democracies where citizens can express openly what they think. Elements of the developed world co-exist with old and traditional elements of societal structures.

The WHO has developed alongside the rest of the world since its establishment in 1948. Its initial budget in 1948 was US$ 5 million. Budget-wise and manpower-wise, the WHO dimension has expanded considerably.[5–8] The currently approved budget for 2020–2021 is US$ 5.84 billion. Due to the pandemic outbreak, additional funds are raised to address COVID-19.[29]

The WHO reflects today's world with its bizarre juxtaposition of modern, archaic, and mixed elements. But it has also acquired a life of its own. Its self-description today is: "Dedicated to the well-being of all people and guided by science, the World Health Organization leads and champions global efforts to give everyone, everywhere an equal chance to live a healthy life."[5] It has acquired a global existence on the internet and is often mentioned in the media. It delivers annually the World Health Report, the worldwide World Health Survey, and runs the World Health Day.

The WHO's objectives are defined in its constitution with a multitude of coordinating and directing tasks across all aspects of international public health. It claims to provide leadership; it engages in partnerships, wants to shape the research agenda, to stimulate generation and dissemination of knowledge, to set norms and standards; to articulate ethical and evidence-based policy options; to provide technical support, catalyze change, build institutional capacity; and to monitor health and health trends.[30] It describes itself as dedicated to the well-being of all people and guided by science.[5]

All this sounds impressive, but does it reflect the real world? Unfortunately, it reflects more the current illusions in academic circles and state-supported TV and radio channels, as well as the self-portrayals of governments and government institutions presented through the diplomatic representation of their governments at the international level.

There is a strong desire in academic and diplomatic circles that the UN and it's WHO should have higher legal, political, and scientific legitimacy than individual countries and science. This illusion project dreams of a world government that in reality does not exist. But the annual budget of more than the US $ 5 billion, 7000 employees, the well-maintained WHO website, and the dreams and illusions of well-off people both in the developed and in the developing world that is projected into the WHO give it also life on its own. Health is a matter in which everyone has their own experience as opposed to official propaganda and indoctrination. Especially where concrete experiences tend to be negative and the hopes for an intervention by the respective state and its public health service are limited from the outset, the existence of an international organization with official legitimacy beyond the individual state allows the WHO to appear like a really functioning supranational authority.

4 The WHO and progress in healthcare

When the WHO was established in 1948, penicillin had just become broadly available. Drug development had just started. In the ensuing decades, modern drugs have increasingly changed our world and the composition of mankind, with many people surviving both in developed and developing countries that would have died only decades earlier. The development of new drugs is expensive. Companies invest considerable funds. They hope to recoup their investment after drug approval by selling the patent-protected drugs until the patents expire, the drugs become generic, and generic copies become available more cheaply. The pharmaceutical industry is part of the industrialized world with corresponding pricing structures. Only a limited proportion of their products reach developing countries. Decisions about drug development are influenced by the potential return on sales as well as the feasibility and cost of its development. By its respective logic, pharmaceutical companies seek the highest feasible price for their new drugs whereas public sector users search for the lowest. But in our increasingly complex and global world, the need for new drugs is constantly evolving, pushed by newly emerging diseases, resistance to antibiotics and chemotherapy, advances in molecular biology, and more. Until the 1980s, there had been little controversy that patent-protected drugs were in general beyond the reach of health services in developing countries. The WHO compiled a list of essential drugs addressed to developing countries that excluded new drugs. But in the late 1980s the WHO published a complementary list of partially patent-protected antibiotics as the only feasible way of dealing with the increasing resistance to antibacterial drugs. Then the HIV/AIDS epidemic led to a public discussion of antiretrovirals and other drugs needed by patients in developing countries.[31]

In 1998, 39 pharmaceutical companies sued the South African government to prevent the implementation of a law designed to facilitate access to AIDS drugs at low cost. The companies accused South Africa, which had the largest population of persons living with HIV/AIDS worldwide, of circumventing patent protections guaranteed by intellectual property rules included in the latest round of world trade agreements. The companies dropped their lawsuit in 2001 after an avalanche of negative publicity. On a public relations level, the decision to sue the South African government was a major disaster for the pharmaceutical industry. These events showed not only the possibilities for coordinated political activism in the era of modern global communications, but also the complex social, economic, and political dimensions of HIV/AIDS treatment in poor countries. South Africa government responses to HIV/AIDS had been inconsistent, controversial, and bizarre. President Thabo Mbeki made completely wrong statements about the causes of HIV/AIDS and questioned the appropriateness and safety of antiretroviral therapy, harming the government's credibility with the international scientific and donor community. In 1997, the government passed the 1997 Medicines and Related Substances Control Act, which became the object of the mentioned legal challenge. The government failed to use the emergency powers that would have made it easier and cheaper to obtain antiretroviral drugs. After having withdrawn their lawsuit, companies began to negotiate sharply reduced prices for some drugs, but the Minister of Health announced that purchasing and distributing antiretrovirals would not soon be a high government priority.[32] In the early 2000s, the South African Treatment Action Campaign (TAC) played a pivotal role in ensuring that pregnant HIV-positive South African women received the antiretroviral drug nevirapine which could help prevent HIV transmission through birth. TAC sued the government, won the case, and thus helped thousands of HIV-positive mothers give birth to healthy, HIV-negative babies, saved many lives, and continues to do so until today.[33]

The Agreement on Trade-Related Aspects of Intellectual Property Rights (TRIPS) is an international legal agreement between all the member nations of the World Trade Organization (WTO). It establishes minimum standards for the regulation by national governments of different forms of intellectual property. The TRIPS agreement introduced intellectual property law into the multilateral trading system for the first time and remains until today. In 2001, developing countries initiated talks that resulted in the WTO Doha Declaration that clarifies the scope of TRIPS, stating for example that TRIPS should be interpreted with the goal to promote access to medicines for all. TRIPS requires WTO members to provide copyright rights to producers of many different types of products.[34] TRIPS allows the issuing of compulsory licenses to market copies of new drugs. The humanitarian intent of this is not in question, but this carries the danger that companies will no longer involve themselves in the future development of innovative drugs.[31]

Pharmaceutical companies sponsor philanthropic projects. Many are coordinated by the WHO. Because of extreme poverty, such efforts are never enough. What action the industry can take, and how much is reasonable without compromising research strategy is a never-ending question. Companies that fail on the market will go bankrupt. Many external critics are unable to see or do not want to see this.[31]

The WHO does not drive drug development. It began as an agency concerned with making low-cost medications available also to the developing world. It has become increasingly a voice for more distribution of modern drugs for a low price to the developing world, appealing to humanitarian feelings. Drug development is entrepreneurial and market-driven. It would be impossible without science, but there is a huge difference between publishing a critical observation, such as Katalin Karikó did on the need to modify messenger RNA (mRNA) to avoid it being attacked and destroyed by the body's immune system,[35] and the development of drugs and vaccines based on this observation. Dr. Karikó discovered a way to configure mRNA so that it slipped past the body's natural defenses. Eventually, this led to the development of two effective mRNA-based COVID-19 vaccines. Dr. Karikó joined the—then—tiny start-up pharmaceutical company BioNTech that then did not even have its own website.[36] Development of drugs and vaccines is done by human beings in pharmaceutical companies. Governments can offer good framework conditions, but development work has to be done by real people. We should not confuse hard development work with government decisions.

Academia, EU governments, and the European Medicines Agency (EMA) dislike the fact that in the end it is an entrepreneurial industry that changes the history of medicine. The WHO, the EU, and the EMA start acting when effective medication is already around. They prefer not to reflect the steps that precede the availability of modern powerful drugs and vaccines.

The WHO expressed its position towards drug discovery and drug development in a 2004 document co-signed by UNICEF, UNDP, the World Bank, and the WHO Special Program for Research and Training in Tropical Diseases (TDR). It outlines that for the diseases within TDR's mandate (leprosy, malaria, tuberculosis, and more) there are no vaccines and no prospects for vaccines becoming available soon. TDR will promote, enable and conduct the development and registration of new drugs for neglected tropical diseases. Furthermore, it describes that the vast majority of drugs available to date have at least one of the following drawbacks: (1) insufficient efficacy or increasing loss thereof, (2) high level of toxicity, (3) inaccessibility, and/or (4) high costs.[37]

In the past, there have been drugs that were detected by academic scientists or by screening traditional medicines. Penicillin was found by Alexander Fleming in the UK. It revolutionized medicine only after Fleming's successors managed during world war II to get support from the same governmental institution that also coordinated the Manhattan project to develop the first atomic bomb.[38] The currently most important drug against malaria, Artemisinin, is extracted from the plant *Artemisia annua* (sweet wormwood), a herb employed in Chinese traditional medicine.[39] But the heyday of finding new drugs through academic institutions is over. Academic researchers do not like to hear this, which is understandable. The WHO TDR works with the world's best-known and most prestigious academic research institutions, including the Swiss Tropical Institute, Basel, Switzerland, and the London School of Hygiene and Tropical Medicine.[37] Nonetheless, drug development is expensive, and in the long term, innovative drugs against infectious diseases require high-tech development, which has become a domain of the pharmaceutical and life science industries. And high-tech drug development is expensive.

5 The WHO in the COVID-19 pandemic

The WHO declared the COVID-19 pandemic in January 2020 a PHEIC. Few countries followed the WHO's call for testing, tracing, and social distancing. By mid-March, COVID-19 had spread around the world.[40] China has achieved remarkable results in suppressing the spread of the pandemic after it denied its existence for the first weeks. But China is an authoritarian state that can suppress the freedoms of its citizens at any time. Such restrictions on personal freedom are not possible in most countries in the free world. Some countries initially slowed the spread of the pandemic by tracing and isolating first infections. Nonetheless, an effective vaccine is needed to overcome the pandemic long-term.

Executive power rests with the respective individual countries. The WHO established a program to deliver testing, protective, and medical supplies for prevention and treatment of COVID-19 to low-income countries, which sounds nice on the surface, but before we can talk about the distribution of vaccines, they need to be developed, approved, and produced. In other words: they must exist physically.

COVID-19 Vaccines Global Access (COVAX) is a worldwide initiative aimed at equitable access to COVID-19 vaccines directed by "Gavi, the Vaccine Alliance" (formerly the Global Alliance for Vaccines and

Immunization, or GAVI), the Coalition for Epidemic Preparedness Innovations (CEPI), and the WHO. It is one of the three pillars of the Access to COVID-19 Tools Accelerator, an initiative begun in April 2020 by the WHO, the European Commission, and the French government as a response to the COVID-19 pandemic. COVAX aims at enabling low-to-middle-income countries equitable access to COVID-19 tests, therapies, and vaccines.[41]

The WHO praised China's public health response to the COVID-19 pandemic, while seeking to maintain a diplomatic balance between the US and China. China's official reports of cases and deaths came late, were always tainted by underreporting, and were of very limited credibility. This was particularly relevant as the pandemic originated in China. Also, the mortality rates reported by Russia were initially much lower compared to other hard-hit countries. The official numbers of COVID-19 deaths published by Russia's federal statistics service sharply contradict the number of deaths reported by Russia's health ministry, reporting an almost three times higher death toll than Russia's health ministry's figure for 2020. Many more countries underreported COVID-19 deaths.[42,43] In other words: the usual theater that we know from all other matters where politics are involved. For an own picture, one must distinguish official propaganda from more trustworthy sources such as peer-reviewed medical journals, leading Western newspapers, or within limitations also Wikipedia. The WHO depends on the payments of the member countries and is therefore not independent.

In the described theater, most European countries and scientists quite understand of the WHO, confusing the desire for an anti-colonial and anti-racist attitude with the need to call things by their names. The former US President Donald Trump criticized the WHO as severely mismanaging and covering up the spread of the coronavirus and finally notified the UN of the US withdrawal from the WHO.[6] While the criticism of the WHO was by and largely correct, it did not represent a great beacon of the Western world's ability to adequately assess the COVID-19 pandemic, considering former president Trump's erratic attitude. Trump's successor Joe Biden canceled the planned withdrawal and announced that the US would resume funding the WHO.

In the end, we can only state that the existence of modern pharmaceutical companies, which work based on an interface between basic research, applied research, and drug approval, are the most important single factor that allows the COVID-19 pandemic to be overcome. We need a critical stance towards the political structures of mankind and their pleasant declarations.

It is impossible and unnecessary to comprehensively analyze all WHO documents and individual contributions. As in all administrations and institutions, there are capable and less capable employees in the

WHO, always within the framework of what the sending countries allow or order their respective institutions. In the following, we analyze a key document that provides insight into the mindset of the WHO and the circles instrumental in its relation to the COVID-19 pandemic.[44]

6 The independent panel for pandemic preparedness and response (IPPR) report

In 2020 the World Health Assembly (WHA) asked the WHO Director-General for a review of the international health response to COVID-19, of lessons learned, and for recommendations for the future. The Director-General asked Ellen Johnson Sirleaf and Helen Clark to convene an independent panel and to report to the WHA in 2021. The report has the promising title "COVID-19: Make it the last pandemic: by The Independent Panel for Pandemic Preparedness and Response"[44] Table 2 lists the members of the independent panel.

TABLE 2   The WHO Independent Panel for Pandemic Preparedness and Response (IPPR).

Co-chairs	(Members, continued)
Helen Clark[a]	Joanne Liu[g]
Ellen Johnson Sirleaf[b]	Precious Matsoso[h]
Members	David Miliband[i]
Mauricio Cárdenas[c]	Thoraya Obaid[j]
Aya Chebbi[d]	Preeti Sudan[k]
Mark Dybul[e]	Ernesto Zedillo[l]
Michel Kazatchkine[f]	Zhong Nanshan[m]

[a] *Prime Minister of New Zealand 1999–2008. The administrator of the UN Development Program 2009–2017*[45].
[b] *President of Liberia 2006–2018. First elected female head of state in Africa.*[46]
[c] *Former Minister of Finance of Colombia; had other ministerial responsibilities before*[47].
[d] *Tunisian diplomat and pan-African and feminist activist. First African Union (AU) Envoy on Youth in 2018. A youngest senior official in the AU history*[48].
[e] *US diplomat, physician, and medical researcher. Was executive director of The Global Fund to Fight AIDS, Tuberculosis and Malaria 2012–2017.*[49]
[f] *French physician and diplomat. Director of The Global Fund to Fight AIDS, Tuberculosis and Malaria 2007–2012. UN Special Envoy for HIV/AIDS in Eastern Europe and Central Asia.*[50]
[g] *Canadian pediatric emergency physician. Professor of Clinical Medicine at McGill University, previous International President of Doctors without borders (MSF)*[51].
[h] *Pulmonologist, president of the Chinese Medical Association 2005–2009. Managed the SARS outbreak, refuting the official line which downplayed its severity. Leading advisor in managing COVID-19.*[52]
[i] *Director general, South Africa Department of Health. Degrees in Law and Ethics, Pharmacy, and more. Was director, public health innovation and intellectual property in the office of the WHO Director General.*[53]
[j] *President of the International Rescue Committee and former British Labour Party politician. Secretary of State for Foreign and Commonwealth Affairs 2007–2010.*[54]
[k] *First Saudi Arabian head of a UN agency (UNFPA). Established the first women's development program in Western Asia. Many further high-level international positions.*[55,56]
[l] *Health Secretary of India 2017–2020. The key strategist in the COVID-19 pandemic.*[57]
[m] *Mexican economist and politician. President of Mexico 1994–2000.*[58]

In their report, the authors describe that they "heard loud and clear that citizens are demanding an end to this pandemic, and that is what they deserve. It is the responsibility of leaders of all countries, as duty bearers, to respond to these demands." They emphasize that a groundswell of opinion is determined to address inequality; that vaccination alone will not end this pandemic, but that it must be combined with testing, contact-tracing, isolation, quarantine, masking, physical distancing, hand hygiene, and more. Vaccine rollout should be scaled up urgently and equitably across the world. They criticize that high-income countries have over 200% population coverage of vaccine doses, while many low and middle-income countries have been shut out of these arrangements. The manufacturing capacity of mRNA and other vaccines should be urgently be built in Africa, Latin America, and other low- and middle-income regions. Agreements on voluntary licensing and technology transfer should be reached.

The report summarizes its demands in six points, shown here in abbreviated form

1. Apply non-pharmaceutical public health measures in all countries. All countries to have an explicit evidence-based strategy to curb COVID-19 transmission.
2. High-income countries to provide to low and middle-income countries 1 billion vaccine doses until Sept 2021; over 2 billion doses by mid-2022 thru COVAX and other mechanisms.
3. G7 countries to commit to providing 60% of the US$ 19 billion in 2021 for vaccines, diagnostics, therapeutics, and strengthening health systems. The rest are to be mobilized from G20 and other higher-income countries.
4. WTO and WHO to convince vaccine-producing countries and manufacturers to agree on voluntary licensing and technology transfer for COVID-19 vaccines. Otherwise, waiver of intellectual property rights under the TRIPS agreement.
5. Production of and access to COVID-19 tests and therapeutics, including oxygen, to be scaled up in low- and middle- income countries. Full funding by the Global Fund's COVID-19 Response Mechanism.
6. WHO to develop a roadmap for short-, medium-, and long-term responses to COVID-19, with clear goals, targets, and milestones to guide and monitor their implementation.

The authors conclude that despite many warnings since the first outbreak of SARS, COVID-19 still took large parts of the world by surprise. The declaration of a PHEIC did not lead to an urgent, coordinated, worldwide response. They emphasize a need for:

- Stronger leadership and better coordination at the national, regional, and international level, a more focused and independent WHO, a Pandemic Treaty, and more.

- Investment in preparedness now, and not when the next crisis hits.
- An improved surveillance system; authority for WHO to publish info and to dispatch expert missions immediately.
- A platform to produce vaccines, diagnostics, therapeutics, and supplies and secure their rapid and equitable delivery.
- Access to financial resources for preparedness and at the onset of a potential pandemic.[44]

7 Assessment of the WHO IPPR recommendations

The IPPR report is completely right in its assessment that the entire world was surprised by the COVID-19 pandemic and that the level of preparedness was utterly inadequate. But it avoids mentioning the suppression of first warnings by a Chinese ophthalmologist, and how the Chinese government is trying to gloss over its role during the pandemic.

The WHO does not represent a world government. The IPPR represents a group of persons who have become known through political and health activism, national elections, or other career mechanisms that the WHA can feel represented in it. The political achievements of the IPPR members and other WHO representatives are in no way called into question here. In the contrary, their resumes are impressive. But here we do not discuss personal achievements, democratic ideals, or political propaganda platforms. We seek to shed light on reality, with a particular focus on the interfaces between health care institutions and political decision-making mechanisms. In this analysis, we must distinguish wishful thinking from reality, even if that will not please many main players.

The IPPR members may have heard many times loud and clear citizens' demand for an end to this pandemic. To report such personal demands should not be part of an official report. Of course, citizens demand an end to the pandemic. To repeat these demands in a report that claims to represent a global healthcare authority is ridiculous. Such demands are wishful thinking. They are irrelevant when we think about practical steps forward. However, they are relevant for politicians who want to be elected, or for officials who try to obtain higher budgets and resources through publicly raised demands.

In the discussion of the COVID-19 pandemic, the division of the world into high-income, medium-income, and low-income countries represents outdated and dusty ideologies of class struggle and the struggle against colonialism and imperialism. The EU in particular, which definitely represents a high-income continent, has proven that idiotic politics are possible despite high resources. Whether the preachers and advocates for world reform like it or not, the COVID-19 pandemic is a challenge that can only be overcome by high-tech measures. In this context, class struggle and anti-colonial demands are hollow and empty. Technically, everyone in the EU and North America could already have been vaccinated. The really crucial question is why this has not been achieved. The answer to that question is less easy than people can imagine who prefer to be seen in television interviews and talk shows.

One key element is the demand for voluntary licensing and technology transfer for COVID-19 vaccines, combined with the threat of a waiver of intellectual property rights under the TRIPS agreement. Developing innovative drugs and vaccines is a risky business. Not all attempts are successful. In retrospect, of course, we know which strategy was successful. But taking away the intellectual property would guarantee that in the future no company would devote itself to developing vaccines, drugs, and diagnostics for the next pandemic.

The IPPRI demands would result in a strengthening of the WHO as a supranational administration. Administrations have the ability to expand and absorb more and more funds. But this would not increase their effectiveness. The use of pseudo-economic jargon that speaks of clear goals, targets, and milestones give the IPPR the appearance of respectability as long as you do not read between the lines and see that all demands amount to the allocation of even more funds.

The WHO's assessment of the COVID-19 pandemic is not a document that provides leadership. Instead, it demands support for the developing world, appealing to humanitarian feelings.

Many people in developed countries have a bad conscience towards developing countries. The WHO serves also as a fig leaf for pretending that something meaningful is happening around the world.

8 Conclusions

There is no doubt about the goodwill of the WHO leadership, apart from those IPPR members and WHO employees who represent authoritarian regimes. These people know exactly what to say officially, what to say off the record, and what better not to say at all, even off the record. They also know exactly what their respective governments expect of them in WHO. Such restrictions apply to all international organizations representing different forms of government. WHO, its staff, and its representatives deserve respect for their work and their goodwill. But the developed world must also learn that, despite respect and diplomacy, it must not take the WHO's claims to leadership at face value.

The WHO IPPR report reflects not only a self-overestimation of WHO, but also a schoolmasterly attitude towards the decision-makers of the developed world. The entire world was indeed much unprepared for the COVID-19 pandemic. But the scientific and technical prerequisites for overcoming the pandemic were in place. The only country that provided true leadership on a technical level was the US through Operation Warp Speed, accelerating the development of effective vaccines to a development time below 1 year. There are many true heroes such as the nurses and medical doctors who sacrificed themselves to care for ill patients, including Li Wenliang, who was summoned and admonished by the Wuhan Police for "making false comments on the Internet about unconfirmed SARS outbreak,"[59,60] and thousands of others.

Nevertheless, we have to distinguish between those heroes who did what they could within the framework of the tools they had at their disposal and the heroes who laid the foundations for overcoming the COVID-19 pandemic permanently. The WHO leaders do not belong to the latter heroes.

The real barriers to overcoming this and other pandemics even faster are not the division of the world into high-income, medium-income, and low-income countries, but the overconfidence of individual academics and of institutions in healthcare, with a resulting resistance to rapid implementation of what technologically is already possible. Hence it is that illusions become real obstacles. The WHO IPPR report is a good example that it would be harmful and counterproductive to give in to current WHO demands. We need a strengthened pharmaceutical industry to be fully prepared for the next pandemics, not a weakened industry.

References

1. Weidling P. *International Health Organisations and Movements, 1918–1939*. Cambridge, UK: Cambridge University Press; 1995.
2. Dubin MD. *Summary of Chapter 4 of Paul Weidling, International Health Organisations and Movements, 1918–1939: The League of Nations Health Organisation*. Cambridge University Press; 2009. https://www.cambridge.org/core/books/abs/international-health-organisations-and-movements-19181939/league-of-nations-health-organisation/0A25D328AD38CFF408610704C3E594F6.
3. WHO. *International Statistical Classification of Diseases and Related Health Problems (ICD)*. https://www.who.int/standards/classifications/classification-of-diseases.
4. Wikipedia. *International Classification of Diseases*. https://en.wikipedia.org/wiki/International_Classification_of_Diseases.
5. *About WHO*. https://www.who.int/about.
6. Wikipedia. *World Health Organisation*. https://en.wikipedia.org/wiki/World_Health_Organization.
7. McCarthy M. A brief history of the World Health Organization. *Lancet*. 2002;360:1111–1112. https://www.thelancet.com/pdfs/journals/lancet/PIIS014067360211244X.pdf.
8. *Brief History of WHO*. https://ccnmtl.columbia.edu/projects/caseconsortium/casestudies/112/casestudy/www/layout/case_id_112_id_776.html.
9. *World Health Assembly (WHA)*. https://www.who.int/about/governance/world-health-assembly.
10. *WHO TDR*. https://tdr.who.int/about-us.
11. Wikipedia. *International Health Regulations* (IHR). https://en.wikipedia.org/wiki/International_Health_Regulations.

12. WHO. *International Health Regulations (IHR)*. https://www.who.int/health-topics/international-health-regulations#tab=tab_1.
13. Mullen L, Potter C, Gostin LO, et al. An analysis of international Health regulations emergency committees and public Health emergency of international concern designations. *BMJ Glob Health*. 2020;5(6), e002502. https://gh.bmj.com/content/bmjgh/5/6/e002502.full.pdf.
14. Wikipedia. *World Economy Forum*. https://en.wikipedia.org/wiki/World_Economic_Forum.
15. *World Economic Forum*. https://www.weforum.org/.
16. *Doctors Without Borders (Medicines Sans Frontieres, MSF)*. www.doctorswithoutborders.org/.
17. UN Refugee Agency. *Rescue of Asylum Seekers from Distress at Sea*; 1985. https://www.unhcr.org/excom/exconc/3ae68c4358/rescue-asylum-seekers-distress-sea.html.
18. Pro Asyl Foundation. Refugees in distress at sea: acting and assisting. In: *Guidance for Skippers and Crews*; 2015. https://www.proasyl.de/material/refugees-in-distress-at-sea-acting-and-assisting/.
19. *The Duty to Rescue Refugees and Migrants at Sea*. University of Oxford, Faculty of Law; 2020. https://www.law.ox.ac.uk/research-subject-groups/centre-criminology/centre-border-criminologies/blog/2020/03/duty-rescue.
20. Achterhold G. *Rescuers in a Dilemma. Europe Discusses the Role of Private Maritime Rescuers in the Mediterranean*. Deutschland.de; 2018. https://www.deutschland.de/en/topic/politics/rescue-of-refugees-in-distress-at-sea-conflict-for-ngos-and-rescuers.
21. Cook P. *Ocean Viking Rescue Operations Expose Libya's Crisis Out at Sea*. Geneva Solutions; 2021. https://genevasolutions.news/peace-humanitarian/ocean-viking-rescue-operations-expose-libya-s-crisis-out-at-sea.
22. Blank SJ, ed. *The Russian Military in Contemporary Perspective*. Carlisle, PA, USA: The United States Army War College; 2019. https://publications.armywarcollege.edu/pubs/3705.pdf.
23. Wikipedia. *List of Military Occupations*. https://en.wikipedia.org/wiki/List_of_military_occupations.
24. *MH17 Ukraine Plane Crash: What We Know*. https://www.bbc.com/news/world-europe-28357880.
25. *Malaysia Airlines Flight 17*. https://en.wikipedia.org/wiki/Malaysia_Airlines_Flight_17.
26. Wikipedia. *Adolf Eichmann*. https://en.wikipedia.org/wiki/Adolf_Eichmann.
27. Wikipedia. *Cold War*. https://en.wikipedia.org/wiki/Cold_War.
28. WHO. *Declaration of Alma-Ata*; 1978. https://www.who.int/publications/almaata_declaration_en.pdf.
29. *WHO Budget*. https://www.who.int/about/accountability/budget.
30. Target Health LCC. *History of the World Health Organization (WHO)*; 2018. https://www.targethealth.com/post/history-of-the-world-health-organization-who.
31. Hardwicke CJ. The World Health Organization and the pharmaceutical industry. Common areas of interest and differing views. *Adverse Drug React Toxicol Rev*. 2002;21(1–2):51–99. https://link.springer.com/content/pdf/10.1007/BF03256183.pdf.
32. Barnard D. In the high court of South Africa, case no. 4138/98: the global politics of access to low-cost AIDS drugs in poor countries. *Kennedy Inst Ethics J*. 2002;12(2):159–174.
33. *The Treatment Action Campaign Court Case on Access to Aids Medication*. South Africa Online; 2021. https://southafrica.co.za/treatment-action-campaign-access-to-aids-medication.html.
34. Wikipedia. *TRIPS Agreement*. https://en.wikipedia.org/wiki/TRIPS_Agreement.
35. Karikó K, Buckstein M, Ni H, et al. Suppression of RNA recognition by toll-like receptors: the impact of nucleoside modification and the evolutionary origin of RNA. *Immunity*. 2005;23(2):165–175. https://www.cell.com/action/showPdf?pii=S1074-7613%2805%2900211-6.

References

36. Bendix A. *BioNTech scientist Katalin Karikó risked her career to develop mRNA vaccines. Americans will start getting her coronavirus shot on Monday.* Insider; 2020. https://www.businessinsider.com/mrna-vaccine-pfizer-moderna-coronavirus-2020-12?r=DE&IR=T.
37. Discovery D, Development D. *UNICEF/UNDP/World Bank/WHO Special Programme for Research & Training in Tropical Diseases (TDR)*; 2004. https://www.who.int/tdr/grants/workplans/en/drug.pdf.
38. Hilts PJ. *Protecting America's Health: The FDA, Business, and One Hundred Years of Regulation.* New York, USA: Alfred A Knopf Publishers; 2003.
39. Artemisinin-based combination therapy. https://www.malariaconsortium.org/pages/112.htm.
40. Maxmen A. *Why did the world's pandemic warning system fail when COVID hit?* Nature News; 2021. https://www.nature.com/articles/d41586-021-00162-4.
41. *Wiki COVAX.* https://en.wikipedia.org/wiki/COVAX.
42. Dyer O. Covid-19: Russia's statistics agency reports much higher death tollthan country's health ministry. *Br Med J.* 2021;372, n440. https://www.bmj.com/content/bmj/372/bmj.n440.full.pdf.
43. Rahmandad H, Lim TY, Sterman J. *Estimating COVID-19 Under-Reporting Across 86 Nations: Implications for Projections and Control*; 2020. Pre-print https://www.medrxiv.org/content/10.1101/2020.06.24.20139451v2.full.pdf.
44. Clark H, Sirleaf EJ, et al. *COVID-19: Make it the Last Pandemic.* The Independent Panel for Pandemic Preparedness & Response; 2021. https://theindependentpanel.org/wp-content/uploads/2021/05/COVID-19-Make-it-the-Last-Pandemic_final.pdf.
45. Wikipedia. *Helen Clark.* https://en.wikipedia.org/wiki/Helen_Clark.
46. Wikipedia. *Ellen Johnson Sirleaf.* https://en.wikipedia.org/wiki/Ellen_Johnson_Sirleaf.
47. Wikipedia. *Mauricio Cardenas Santamaria.* https://en.wikipedia.org/wiki/Mauricio_C%C3%A1rdenas_Santamar%C3%ADa.
48. Wikipedia. *Aya Chebbi.* https://en.wikipedia.org/wiki/Aya_Chebbi.
49. Wikipedia. *Mark Dybul.* https://en.wikipedia.org/wiki/Mark_R._Dybul.
50. Wikipedia. *Michel Kazatchkine.* https://en.wikipedia.org/wiki/Michel_Kazatchkine.
51. Wikipedia. *Joanne Liu.* https://en.wikipedia.org/wiki/Joanne_Liu.
52. Wikipedia. *Zhong Nanshan.* https://en.wikipedia.org/wiki/Zhong_Nanshan.
53. Wikimzansi: *Precious Matsoso.* https://wikimzansi.com/malebona-precious-matsoso/.
54. Wikipedia. *David Miliband.* https://en.wikipedia.org/wiki/David_Miliband.
55. Wikipedia. *Thoraya Obaid.* https://de.wikipedia.org/wiki/Soraya_Obaid.
56. https://www.unssc.org/about-unssc/speakers-and-collaborators/dr-thoraya-obaid/.
57. Wikipedia. *Preeti Sudan.* https://en.wikipedia.org/wiki/Preeti_Sudan.
58. Wikipedia. *Ernesto Zedillo.* https://en.wikipedia.org/wiki/Ernesto_Zedillo.
59. Wikipedia. *Li Wenliang.* https://en.wikipedia.org/wiki/Li_Wenliang.
60. Hegarty S. *The Chinese Doctor Who Tried to Warn Others About Coronavirus.* BBC World Asia China; 2020. https://www.bbc.com/news/world-asia-china-51364382.

CHAPTER 9

Low-tech and high-tech challenges. Accidents and disasters. Technical and scientific progress and its perception by science and the public

1 Introduction

In developed countries we are used to seeing most things work routinely, starting with having running water, electricity, education, medicine, medical care, hospitals, public transport, and much more. We are also used to many former common environmental sins having stopped, such as throwing away empty beer cans, wine bottles, plastic bags, and more in the great outdoors. We also see that in other countries to which we go on temporary vacation, these things are often less perfectly regulated. To a certain extent, we like the improvisation ability of the people who live in these countries and are used to getting on with little.

Governments should prevent disasters. If they occur nevertheless, governments should help to keep damage to a minimum and to help repair them. The COVID-19 pandemic is economically, medically, and scientifically a disaster of unprecedented proportions, within the limitations made in the previous chapters. Here we discuss past and present accidents and disasters that played prominent roles in public discussions. Some can or could be seen with the naked eye or on a TV screen, such as the crash of an airplane or the explosion of a spaceship. Others, such as the causes and outcomes of chronic poisoning, are less visible. It has often taken years, sometimes decades, centuries, or even millennia to find the causes of disasters. But knowing the causes is often not enough. When a company poisons the environment it usually does not do so out of sheer malice, but because the cheapest way to get rid of waste products is to dispose of them in the sea, in rivers and lakes, or in filling pits. As long as the extent of such disposal was historically limited, the consequences often remained limited and local. With increasing industrialization and globalization, the extent of environmental pollution has increased, as have the dimensions of the damage it can cause, and the level of public perception of such environmental damage. Science plays a key role in the intellectual processing of our world. The history of accidents and disasters is also a history of their defense by naïve, biased, or even bought scientists, their processing by upright and indomitable scientists, of acceptance by the general population, and of the legal, political, and administrative rules that changed as a result of often initially controversial discussions. Many disasters also show how governments and institutions in different countries, regions, and continents react at different speeds, and how some institutions and governments react fast, while others are almost immune to the emergence of new knowledge.

We begin with man-made disasters that have been discussed broadly in the media. Eventually, technical and organizational reasons were found for all of them. But behind technical and organizational errors are often group psychological and cultural factors and conflicts of interest at the interface between different areas of influence. These forces often resist correction. We then discuss some environmental disasters; the largest industrial accident in history; how tobacco smoking and lead poisoning were gradually recognized as massively damaging health, and how for decades public knowledge of the involved dangers was prevented through skillful pseudoscientific activism, bribery of scientists, and other means. Last not least, we discuss an appealingly low-tech challenge, the introduction of "modern," industrially produced cattle feed from animal protein, intended to reduce the cost of beef production. The consequence was first a frequent illness of animals—bovine spongiform encephalopathy (BSE, mad cow disease), which then turned out to be transmittable to humans. But to understand how flawed food led to a contagious disease in animals, and that the resulting animal disease was transmissible to humans represented a high-tech challenge. Finally, we discuss parallels in these disasters and how they can help us to understand the dynamics that influence the processing and control of the COVID-19 pandemic.

2 Groupthink

When in 1941 the Japanese armed forces attacked the US naval base at Pearl Harbor, Hawaii, the US was still neutral in World War II. The US had intercepted messages and was aware that Japan was arming itself for an offensive attack somewhere. Washington warned the officers stationed at Pearl Harbor, but their warning was not taken seriously.[1-4] Groupthink is a psychological phenomenon that occurs when in a group of decision-makers the desire for harmony or conformity results in irrational or dysfunctional decisions, because the group reaches a consensus without sufficient critical evaluation. The term was coined in 1952 but most initial research was conducted by Irving L Janis who applied small-group analysis to the explanation of policy fiascoes, including the mentioned US military's reluctance to take warnings of a potential Japanese attack seriously. The US armed forces on Pearl Harbor were completely surprised and suffered heavy losses.[1,2,5,6]

In 1961, at the height of the Cold War, Cuban exiles tried to invade Cuba to overthrow Fidel Castro. The operation, which had been covertly financed by the US government, failed miserably. The operation, which was later called by the place of the attempted landing "Bay of Pigs" fiasco, became another epitome of wrong decisions that can occur if key people are too optimistic and rigid in their belief that a mission will succeed, being unknowingly biased by wishful thinking.[1,2,7]

Although the Japanese attack on Pearl Harbor was successful in the short run, there were severe flaws in their war strategy in general, as well as tactical aspects of the attack on Pearl Harbor. The attack neglected the considerable oil reserves in and around Pearl Harbor and did not destroy the repair shops of the naval base. It brought the US government into the war, unified the US population, and led ultimately to the defeat of Japan.[5,6] As most battleships were temporarily out of order, the US now relied on aircraft carriers and submarines. Pearl Harbor's repair shops and fuel depots allowed to continue logistical support to the navy's operations. Submarines immobilized the Japanese navy's heavy ships and crippled the importation of oil and raw materials.[5,6] One could argue that also the leadership of the Japanese Armed Forces and the Japanese government made key decisions under groupthink. This is the downside of psychological explanations that explain everything in hindsight. They can be applied retrospectively to anything that went wrong, which is easy but can result in a know-it-all attitude. Nevertheless, the term "groupthink" has become one of several key elements that can help to analyze flawed decision-making.

3 US space shuttle disasters

In 1982, the US space shuttle Challenger broke apart shortly after beginning its flight, killing all seven crew members. A special commission found that the organizational culture of the National Aeronautics and Space Administration (NASA) had massively contributed to the accident. NASA managers had known that the design of the solid rocket boosters (additional rockets that give additional lift during launch) had a fundamental flaw, but had failed to address this properly. The boosters were transported to the launch site in a disassembled state and were then put together. The sealing rings that should seal the originally disassembled parts did not seal properly, especially when it was too cold outside, which was the case when the Challenger started. This failure caused a breach and allowed pressurized burning gas from within the booster's rocket motor to escape and damage other parts of the vehicle. The entire space shuttle broke apart, which was observed on television by millions in the US and worldwide. NASA managers had also disregarded warnings from engineers about the dangers of launching at low temperatures and did not report these concerns adequately to their superiors. Richard Feynman, a theoretical physicist, and Nobel Prize winner, was a member of the commission that investigated the disaster. He contacts most other commission members, he spoke face-to-face also with the involved engineers, and not only with NASA managers and the other members of the special commission. He became rather critical of flaws in NASA's safety culture. To have his findings included in the final report, he had to threaten to remove his name from the commission report. He outlined that the estimates of reliability offered by NASA management were wildly unrealistic, differing as much as a thousandfold from the estimates of working engineers. He concluded that "for a successful technology, reality must take precedence over public relations, for nature cannot be fooled."[8]

Sociologist Diane Vaughan coined the term "normalization of deviance" when she reviewed the Challenger disaster, noting that the disaster's root cause was related to the repeated choice to fly the shuttle despite known flaws. She described this phenomenon as occurring when people within an organization become so insensitive to deviant practice that it no longer feels wrong. Insensitivity occurs insidiously, sometimes over years, because disasters do not happen until other critical factors line up.[9,10] Also the term "groupthink" is often used to explain the flawed decisions of the NASA management.[11,12]

In 2003, the Columbia space shuttle disintegrated as it reentered the atmosphere after docking with the international space station for a while. All seven crew members were killed. During its launch, a piece of the foam insulation had broken off from the shuttle external tank and had struck the shuttle's left-wing. Similar sheddings had occurred during previous launches. This time some engineers suspected more serious damage. However, NASA managers limited the investigation before re-entry, believing that the crew would not be able to fix the problem in space anyway. When Columbia re-entered the atmosphere, hot atmospheric gases penetrated the heat shield and destroyed the wing structure. The shuttle became unstable and broke apart. The Columbia accident investigation report confirmed the technical causes of the disaster, but also addressed critically its underlying organizational and cultural causes, and concluded that the organizational structure and processes were flawed. The shuttle program manager was responsible both for safe and timely launches as well as acceptable costs, i.e., for potentially conflicting goals. NASA had accepted deviations from design criteria as normal when they happened on several flights. In hindsight, there would have been various options to repair the damage and/or to allow the astronauts to return to earth with other means of transportation.[13]

4 Boeing 737 MAX crashes

Airplanes are highly complex, high-tech devices. When a plane crashes, most passengers die. The construction and licensing of planes are strictly controlled, in the USA by the Federal Aviation Administration (FAA). The Boeing 737 is the world's most popular family of jet-powered commercial aircraft, now being manufactured in the third and fourth generations. It has hardly changed its external appearance over time since the release of its first generation in 1967. The new Boeing 737 MAX was intended to take advantage of a new generation of jet engines to save fuel. Boeing provided it with the computer software "Maneuvering Characteristics Augmentation System" (MCAS) to enhance its flight stability, as the MAX had to deal with the aerodynamic effect of the larger, heavier, and more powerful engines. MCAS could be triggered into action by one of two sensors measuring if the plane was headed up or down. It did not require both sensors to report the same measuring. A special instrument that should let pilots know when the two sensors disagreed was inoperable in most delivered 737 MAX jets. Boeing was aware of this but did not tell airlines. Furthermore, the 737 MAX's chief engineer approved MCAS without fully understanding it.[14] In the 1960s and 1970s, the aircraft manufacturing industry had been centered in the US, and the FAA was the leader in setting standards. Thereafter, other countries and companies such as the European Airbus grew in stature.[15]

The MCAS pitched the airplane's nose down under certain circumstances. The manufacturer had not entered a description of MCAS into the flight manuals. It took two crashes of almost brand new aircraft with together 346 deaths until the Boeing 737 MAX was grounded worldwide. The US FAA resisted grounding longer than many other aircraft regulators worldwide.[16] A US preliminary congressional report claimed that the development of the 737 MAX was marred by technical design failures, lack of transparency with regulators and customers, and obfuscated information about the operation of the aircraft. The report also sharply criticized the FAA which failed to identify key safety issues.[17] There had been too much mutual trust in the relationship between Boeing and the FAA.

5 Nuclear plant meltdown in Fukushima, Japan, after a tsunami

Modern society consumes a lot of energy for production, for cars, airplanes, ships, air conditioners in summer, and heating in winter, and more. The traditional method of transportation depended first on animals for land transport, and on wind for sea transport; then on the burning of coal; then on the burning of oil. Today we are transitioning towards the direct use of electricity for transportation, heating, cooling, and more. While burning coal, oil and gas produces a lot of CO_2, once electricity is present it does not generate waste products. But there are many prejudices against the use of atomic energy, although today we can control its technical challenges. Atomic energy arouses several associations. One is the dropping of two atomic bombs on Hiroshima and Nagasaki during World War II. The other is the fear of meltdown, as it happened in Chernobyl in Ukraine, and in Fukushima in Japan.

Nuclear bombs helped to end World War II, in which Japan committed unimaginable atrocities, comparable in magnitude to the crimes of the Nazis and the Soviet Union.[18] The atomic bombs were part of the warfare. War is inherently cruel. Nevertheless, the memory of atomic bombs still overshadows the peaceful use of atomic energy. Modern technology is complex, and when things go wrong, the damage is enormous. But that is no reason to forego modern technology.

In 2011, a strong undersea mega-thrust earthquake occurred east of Japan, triggering a strong tsunami.[19] On detecting the earthquake, the reactors of the Fukushima Daiichi Nuclear Power Plant, run by the Tokyo Electric Power Company (TEPCO), automatically shut down the power-generating nuclear fission reactions. But electricity was still needed to power the pumps that circulated coolant through the reactors. When the electricity supply failed, the emergency generators started. Circulation of coolant is vital to remove residual decay heat, which continues even when fission has ceased. But the earthquake had also generated a tsunami that swept over the plant's seawall and flooded the lower parts of the reactors. This caused the failure of the emergency generators, and the loss of power. When the emergency batteries had been exhausted and no new emergency generators had been flown in, the loss of reactor core cooling led to nuclear meltdowns, hydrogen explosions, and radioactive contamination of three of the four reactors. The flooding of the emergency generators could have been avoided by constructing them on elevated ground, which would not have been a high-tech challenge. Earthquakes and tsunamis occur relatively often in and around Japan. TEPCO had failed to meet basic precautions including risk assessment, preparing for containing collateral damage, and evacuation plans. The Japanese Ministry of Economy, Trade, and Industry, which should have supervised TEPCO, had an inherent conflict of interest as it was both in charge of regulating the nuclear power industry and promoting the use of nuclear power. The relationship between the supervising agency and TEPCO was too cozy. A large area had to be evacuated, large amounts of contaminated water were released into the ocean, and the area is still struggling. Warnings had been ignored.[20]

Why was the relationship between the Ministry of Economy, Trade, and Industry and TEPCO so cozy? It was partly due to Japanese tradition with a very high level of reverence for institutions. It collided with the need to control the nuclear power plants with impartial severity. Top officials in ministries are politically appointed. Modern technology needs strict controls. Mankind can control the risks of nuclear plants, provided that human weaknesses in control are eliminated as much as possible.

6 Lead poisoning

Lead is a normal constituent of the earth's crust, with trace amounts in soil, water, and everywhere in nature. Undisturbed, lead is practically immobile. However, once mined and turned into artificial products, it is dispersed throughout the environment and shows highly toxic characteristics. In the course of human civilization, lead has become the most widely scattered toxic metal worldwide. Furthermore, lead has a long environmental persistence and never loses its toxic potential if ingested.[21]

The nerve system is most sensitive to lead poisoning. Symptoms of acute poisoning include abdominal pain, constipation, headaches, irritability, memory problems, infertility, tingling in hands and feet, reduced intellectual disability, and behavioral problems. Not all poisonous effects resolve later. Severe cases can lead to anemia, headache, convulsions, muscle weakness, delirium, coma, or death. Chronic poisoning results often in fatigue, sleep disturbance, headache, irritability, slurred speech, stupor, ataxia, convulsions, anemia, and renal failure.[22–27]

It is even worse when children are exposed to chronic lead intoxication.[23] Long-term effects can include brain damage, anemia, impaired kidney function, muscle weakness, learning disabilities, attention deficit disorders, impaired fine and gross motor skills, and more. Adults may have a slow accumulation of lead in the bloodstream and its long-term effects may be harder to diagnose. Lead has a cumulative effect and may impact cognitive function over time as well as lead to central and peripheral damage of the nerve system, anemia, decreased fertility, and abnormal menstrual cycles.[28]

Lead has been mined for several 1000 years. In ancient Rome, it was used extensively in the form of cooking utensils and pots, wine urns, plumbing, to line aqueducts, vessels to concentrate grape juice and to store wine, for makeup, and more. It was used to prevent bacteria from turning wine into vinegar—today's sulfates were not yet known. The Romans shipped wines all over their empire. To preserve and "improve" wine in ancient Rome a sweet concentrate of grape juice was used, called Sapa, produced by boiling down unfermented grape juice into a concentrate in pots made of lead, resulting in a sweet fluid that contained toxic levels of lead by today's standards. Sapa was also used as a sweetening agent in many dishes. These days, the measurement of lead content in fluids was unknown. The ancient literature described acute lead intoxication, but chronic lead intoxication was recognized only in modern times.[21]

The high exposure to lead may even have contributed to the decline of the Roman Empire due to its neurotoxicity that primarily affected the Roman upper class. We know from the literature the strange behavior of many Roman leaders and the upper class in general, but until the recent past, the connection between chronic lead intoxication has not been made. While these days, lead poisoning affected pre-dominantly the affluent strata of society, it became during the last centuries more and more an affliction of the poorer communities.[21,29,30]

Industrially produced lead and its compounds gained more use at the beginning of industrialization in pottery, piping, shipbuilding, window making, arms production, pigments, and book printing. Lead poisoning became a plague in Europe and later in America during the 15th, 16th, 17th, and 18th centuries. In antiquity, acute lead poisoning had been mentioned in the literature. In the 16th century, it reappeared in the medical literature. For example, Paracelsus described the "miner's disease." In the 19th and the first half of the 20th-century women working in lead industries knew well that lead caused abortions. Lead compounds were widely used to induce illegal abortions. Women industrially exposed to lead were more likely to be sterile, and if they became pregnant, the risk of stillbirth was much higher.[24]

In modern times, two additional major avenues of lead pollution evolved.

1. Lead was used as the primary pigment in house paint since the 1880s.[31] Lead-based paint became popular during colonial times for use on interiors and exteriors of homes, partially due to its durability. In the US, the peak of lead paint use was in the 19th century. Most homes built before 1978 contain traces of lead-based paints. Governments in Europe became increasingly concerned with the health risks of lead paint, and the lead white color was gradually replaced by zinc and titanium white instead. The US reacted later, compared to other countries. In 1971, the US Congress banned the use of lead-based paints in any newly built residential or commercial buildings, if they were constructed using federal funding or assistance. In 1977, the US Consumer Product Safety Commission banned the use of lead-based paint in residential and

public properties, as well as the use of lead paint in toys and furniture. The Lead-Based Paint Disclosure Regulation, put in place by the US Environmental Protection Agency (EPA) and the Department of Housing and Urban Development requires owners of homes built in 1978 or earlier to disclose the presence of lead paint to potential buyers or renters.[32]

To this day, deteriorating lead-based paint is a source of lead poisoning in American children, who are usually exposed as a result of eating chips of peeling or flaking house paint or the lead-contaminated soil or dust near lead-painted surfaces or near busy roads where decades of leaded gasoline use contaminated the soil. US President George Bush signed the Residential Lead-Based Paint Hazard Reduction Act in 1992, declaring the danger from lead-based paint a national crisis. Lead does not decompose into harmless substances over time. It still lurks in the flaking, chipping paint and lead-soldered pipes of aging houses. In 2000, the US Department of Housing and Urban Development estimated that about 38 million houses and apartments built before 1978 contain some lead paint.[25]

2. The anti-knock power of tetraethyl lead (TEL) suspended in gasoline was discovered in 1921 at the General Motors (GM) research laboratories. Charles Kettering, GM vice president for research, had engaged in a broad search for antiknock fuel additives to improve engine compression and power. Lead was not the only or even technically the best additive, but it was cheap. Kettering gave leaded gasoline the name "ethyl," which confused the product with ethyl alcohol, another fuel additive widely used in high compression racing engines and anti-knock blends with gasoline. Leaded gasoline was portrayed as a breakthrough. It was introduced on the market in 1923. Motorists liked the extra boost the gasoline gave. Some scientists and public health advocates objected to its introduction, but they could not block the industry's use of lead in gasoline in the 1920s. Its supporters characterized leaded gasoline as a "gift of God," necessary to the functioning of modern civilization. It took two scientists from outside the usual disciplinary constraints to challenge industry: geochemist Clair Cameron Patterson, who exposed flaws in the scientific methods of justifying leaded gasoline,[33] and Herbert Needleman, a psychiatrist whose epidemiologic studies correlated higher lead levels with worse school performance and lower IQ levels in children.[34] Needleman found that teeth made better markers of past lead exposure than blood samples. He collected teeth from 2500 primary school children. After controlling for other confounding variables such as socioeconomic status, they found that as lead levels increased, all measures of school performance decreased significantly.[35]

Shortly after TEL manufacture began in 1923, workers at all plants began to become psychotic and die. A moratorium on TEL production was put into place. At a conference in 1925, the advocates of lead argued that the use of leaded gasoline raised public health issues that were novel. Robert Kehoe provided a logic that would resolve the controversy. He offered to discontinue the sale of leaded gasoline if an actual danger could be shown. But if such a danger could not be shown "based on facts," a product this economically beneficial should not be "thrown into the discard based on opinions." The Kehoe principle, which claims that the absence of evidence of risk is evidence of the absence of risk,[36] dominated the debate on leaded gasoline for the next half-century.

The Kehoe principle placed the burden of proof on the opponents of TEL. They would have to prove that the use of TEL was unsafe. If this proof was inconclusive, then TEL should be allowed. The Kehoe principle also put the interest in protecting public health against the economic benefits of TEL. A precautionary principle would have required proof that TEL was safe before it could be used. With the large investments by industry, the social and economic climate of the time, and the belief in progress, the outcome of the 1925 conference was preordained. The TEL moratorium was lifted in 1926, TEL production resumed, and soon leaded gasoline was commonly used.[29,37,38]

The industry needed to fund and control the research in lead toxicity, done through the Kettering Laboratory under Kehoe's direction. Kehoe's laboratory dominated the scene for decades, attesting to the safety of leaded gasoline and deconstructing any criticism. The credibility of Kehoe's research was bolstered for decades by the support of the US Public Health Service and the American Medical Association.[37]

Between 1926 and 1965, the prevailing consensus was that lead toxicity occurred only at high levels of exposure and that lead in the atmosphere was harmless. Most of the data on lead toxicity was issued from a single source, the Kettering Laboratory in Cincinnati. Finally, in 1965, Kehoe's monopoly on lead science was challenged. Patterson's measurements of the isotopic ratios of minerals showed that the long-held belief that the earth was 3 billion years old was wildly wrong. Now he placed its age at 4.5 billion years. Initially, he swam against a tidal wave of orthodox scientific opinion. Finally, his findings were confirmed, his skeptics silenced, and the geology textbooks revised.[29]

Patterson employed extraordinary measures to avoid contamination while collecting and analyzing his specimens. As a result, the isotope ratios were vastly more accurate than those of earlier workers. He observed that the lead levels in soil and ice were much higher than expected based on natural fluxes and realized that human activity had severely raised environmental levels of lead. He showed that technological activity had raised the modern human body's lead burdens to levels 600 times that of our pretechnologic ancients. Kehoe and his partisans had commonly referred to average population values as "normal," conveying some of the meaning "natural." Patterson showed that because a certain level of lead was now commonplace did not mean it was without harm. In the late 1960s, the question of "silent" lead poisoning drew the attention of civil rights and anti-poverty movements, urban advocates, and environmentalists. In 1970, Congress passed the Clean Air Act, directing the EPA to name each pollutant known to be dangerous and widespread and within 2 years to issue a standard that defined a safe level of exposure. In 1975, US car manufacturers began equipping new model cars with catalytic converters designed to reduce pollution and run only on unleaded fuel. By 1986, the major phase-out of leaded gas in the US was complete. As in many other areas, also regarding the environmental lead poisoning, the European Union (EU) caught up much later, finally banning leaded gasoline in 2000.[25,39]

The payoff for taking lead out of gasoline exceeded predictions. Lead levels in children's and adults' blood continued to drop in direct relationship to the reduction in lead in gasoline. The average American child's blood lead level in 1976 was 13.7 µg/dL. In 1991 it was 3.2 µg/dL. In 1988 the Government estimated that 3–4 million American children had blood lead levels greater than 15 µg/dL, the level than assumed to be toxic. Six years later, in 1994, it was estimated that "only" 600,000 children had blood lead levels in that range. The removal of lead from gasoline spared millions of children from growing up with hazardous lead concentrations in their bodies.[29]

By the early 21st century, it was widely accepted that blood lead levels over 10 µg/dL caused diminutions in the intelligence quotient (IQ) and caused several neuro-cognitive problems. Over the past 20 years a voluminous literature has consistently identified problems, such as increased rates of attention deficit hyperactivity disease (ADHD), impulsivity, disruptive and violent behavior, poor executive functioning and short-term memory, first at levels below 10 µg/dL, then below 5 µg/dL, and now at the very lowest measurable levels of children's blood lead.[31]

7 Love Canal/Blackcreek village

Love Canal is a neighborhood in Niagara Falls, New York. In 1890, William Love, an ambitious entrepreneur, planned to construct an urban community of parks and residences on the banks of Lake Ontario, believing it would serve the area's burgeoning industries with much-needed hydroelectricity, and named the project "Model City, New York." His plan also incorporated a shipping lane that would bypass Niagara Falls. In 1894, work on the canal began. Steel companies and other manufacturers expressed interest in opening plants. Love began having a canal dug and built a few streets and houses. But only 1.6 km of the canal was dug, about 15 m wide and 3–12 m deep, stretching northward from the Niagara River. Industry and tourism increased steadily throughout the first half of the 20th century due to a great demand for industrial products and the increased mobility of people.[40] Eventually, Love's canal was purchased by Hooker Chemicals, which produced chemicals ranging from caustic soda to synthetic pesticides, polyester resins, and polyvinyl chloride (PVC). Hooker Chemicals dumped accruing waste in Love Canal. Once it was filled, it was covered with earth and purchased by the city of Niagara Falls. Love Canal first became part of the suburban American dream, but then a toxic nightmare for its residents. After initially ignoring health problems, miscarriages, and birth defects, the mid-1970s saw officials acknowledge its toxic nature. In 1978, the US president declared Love Canal a national emergency. Funds were granted for the evacuation of a limited number of families and remediation work. US legislators created a superfund for toxic remediation, new toxic waste guidelines, and a national inventory and priorities list for toxic remediation. The superfund was reauthorized in 1986 and included the US' first right-to-know provisions for chemical substances. Love Canal was renamed Blackcreek Village. It does not celebrate grassroots activism. However, with its residents recently taking legal action over birth defects and other health problems, battles over Love Canal's meaning and remediation continued.[41]

8 Mercury poisoning

Mercury is a metal that is ubiquitous in the environment in different chemical forms. In Japan, inorganic mercuric sulfate was used as an industrial catalyst for acetaldehyde synthesis and was discharged into the sea by an industrial plant near the fishing village of Minamata in the 1950s and later also into the Agano River in Niigata in the 1960s. Chronic exposure to mercury in high concentrations is poisonous for the central nervous system and causes the Minamata disease, one of the first and most serious cases of diseases resulting from environmental contamination caused by wastewater discharge from an industrial plant. Methylmercury (MeHg) contained in the wastewater contaminated marine life in the surrounding waters and poisoned those who ingested the affected fish and seafood. During the 1950s, people had begun to see strange phenomena around Minamata Bay. Fish rotated continuously and floated belly-up to the surface, shellfish opened and decomposed, and birds fell while in flight. Cats suffered from excessive salivation and manifested general convulsions, we're unable to walk straight and often collapsed dead. Many drowned in the sea. Eventually, cats were no longer seen in the area.[42–46]

The Minamata Convention on Mercury is an international treaty to protect man from emissions and releases of mercury and mercury compounds. It was approved by delegates representing close to 140 countries in 2013 in Geneva, Switzerland, and adopted and signed later that year at a diplomatic conference in Japan.[47,48]

9 Bhopal

In December of 1984 several 1000 people died within a few hours, many thousands more succumbed later, and well over 100,000 were permanently disabled in Bhopal, India. The disaster was caused was the release of 41 metric tons of volatile methyl isocyanate, a dangerous chemical that had come in contact with water. Ensuing media coverage showed striking contrasts between Union Carbide's Bhopal plant and another Carbide plant making the same pesticide in West Virginia, USA. These double standards included numerous vital aspects of design and operation, compounded by management cutbacks in staffing, worker pay, worker training, and maintenance of functional process safeguards. The worldwide public reaction to the Bhopal disaster led major chemical corporations to issue global corporate policy statements in the late 1980s that said in various ways, "we have no double standards."[49]

10 Tobacco smoking

Tobacco use and exposure to secondhand smoking are the leading preventable causes of morbidity and mortality worldwide, killing more people than tuberculosis, HIV, and malaria combined. More than 1 billion people will die of tobacco-related deaths this century and most adults who smoke started before they were 18 years of age.[31]

An attractive image of the tobacco smoker was in past centuries associated with maturity and glamour, conveying the message that after overcoming the initial repulsion, smoking offers pleasurable effects. Thereafter, addiction becomes the main reason for not stopping.[50]

With the development of manufactured cigarettes, tobacco became a product of mass consumption. Sales increased further with the introduction of a drying process with high temperatures in the 19th century. The tobacco industry had a central role in promoting a huge number of smokers worldwide. Advertising and publicity strategies were very successful. The presence of doctors and nurses in advertisements that promoted smoking worked well for a while. Also, movies played a major role.[50]

The Tobacco Industry Research Committee (TIRC) was founded in 1953 in response to evidence linking tobacco smoking to lung cancer, heart disease, and more. One of TIRC's first acts was to publish the "Frank Statement," marking the start of the prolific use of doubt as an effective tactic to prevent smokers from quitting and to protect them from litigation. Despite the TIRC mission statement to aid and assist research into tobacco use and health, the TIRC rarely conducted or supported research that might reveal a link between smoking and cancer.[51] TIRC changed its name in 1964 to "Council for Tobacco Research" (CTR). TIRC and CTR spent a lot of money on advertising and very little on actual scientific studies.[52] The "Tobacco Institute" was founded in 1958 as a trade association to supplement the work of the TIRC, which later became the Council for Tobacco Research. The TIRC work had been limited to attacking scientific studies that put tobacco in a bad light. In contradistinction, the Tobacco Institute had a broader mission and published "good news" about tobacco and attacked scientific studies critical of tobacco, more by casting doubt on them rather than by rebutting them directly. It also lobbied the US Congress. By 1978 the Tobacco Institute had 70 lobbyists. Senator Ted Kennedy said in 1979, "Dollar for dollar they're probably the most effective lobby on Capitol Hill."[53] The Center for Indoor Air Research (CIAR) was a nonprofit organization, established by three major American tobacco companies in 1988. From 1989 to 1999, the CIAR funded at > 200 published studies, thus establishing itself as a buffer between the tobacco industry and scientists. Many scientists who published research favorable to the industry's position on environmental tobacco smoke accepted funds from the CIAR, but preferred not to accept funds directly from the tobacco industry. The majority of peer-reviewed studies funded by the CIAR served to divert attention from the health effects of smoking, focusing instead on other indoor air toxins. It was simply a front for the tobacco industry.[54]

Until 1994, over 800 private claims were brought against tobacco companies in state courts across the country, asserting negligent manufacture, negligent advertising, fraud, and violation of various state consumer protection statutes. The tobacco companies were successful against these lawsuits. When scientific evidence mounted in the 1980s, tobacco companies argued with contributory negligence, claiming that adverse health effects were previously unknown or lacked substantial credibility.[55]

Early court decisions favored the tobacco companies, but the litigation changed the public perception which now saw the tobacco industry increasingly as intentionally selling products it knew to be harmful. Gradually, the addictive characteristics of nicotine were discovered and it became known that the tobacco companies knew it. Company documents emerged through the legal discovery process or by company whistleblowers, confirming that the companies knew the addictive characteristics of tobacco and welcomed them to maintain their customer base.[56]

Finally, in 1998 the Master Settlement Agreement (MSA) between the state attorneys general and the major US tobacco companies was reached, banning the paid placement of tobacco brands in movies. The settlement dissolved the Tobacco Institute, the CIAR, and the CTR. Before the MSA was finalized, four US states had already reached separate agreements with the cigarette industry. In addition to the $36.8 billion in settlements for these four states, the MSA settlement in the remaining states was $206 billion. The combined total of payments agreed up to all 50 states over the first 25 years was $243 billion, plus $7 billion in additional payments, including a foundation, antismoking education, and enforcement, making a total of about $250 billion.[50,55–60]

Financial settlements in liability cases usually involve either lump sums or structured payments to the claimant, made by the defendant. The MSA imposed fees based on cigarette sales. These fees were tantamount to sales taxes.[60]

The purpose of the MSA was to prevent cigarette smoking and compensate for health expenses incurred in the treatment of tobacco smoking-related illnesses. Twelve years after the settlement, it is clear that MSA monies have been gravely diverted from tobacco prevention and cessation programs to balance budgets.[61]

Since 1980, large reductions have been observed in the percentage of daily smoking at the global level for both men and women. However, paradoxically, as a result of absolute population growth, the total number of smokers has further increased. The pressure in favor of smoking bans in public places, and its consequent ramifications and influence on people's perceptions and attitudes toward smoking have created a feeling of discrimination against smokers. This is very different from the feeling of being a glamorous smoker during most of the previous century. There has been a gradual change in the consumption pattern of cigarettes, with a progressive reduction of daily use in developed countries. Nicotine has some positive cognitive effects and has none of the adverse behavioral effects of other drugs. Now there is an anti-smoking social pressure, and a stigma associated with smoking and the inability to give it up.[50]

When tobacco smoking could no longer be publicly defended, the AMA tried to brush under the rug the fact that it had accepted grants from the tobacco industry for many years.[62,63]

11 BSE and Jacob-Creutzfeld-disease

Prions are misfolded proteins that can transmit their misfolded shape onto normal variants of the same protein. They cause several fatal and transmissible neurodegenerative diseases in animals and humans. Somehow, the abnormal three-dimensional structure of proteins confers infectious properties, collapsing nearby protein molecules into the same shape. The term "prion" stands for "proteinaceous infectious particle."

The role of a protein as an infectious agent stands in contrast to all other known infectious agents such as viruses, bacteria, fungi, and parasites.[64,65] Bovine spongiform encephalopathy (BSE), broadly known as "mad cow disease," is caused by prions and became well known worldwide when it was found that the disease can be transmitted from cattle to humans through the consumption of affected meat.[66,67] In humans, it is called "variant Creutzfeldt–Jakob disease" (vCJD). Creutzfeldt–Jakob disease (CJD) is a lethal human prion disease, which causes progressive cognitive and motor dysfunction. Most affected individuals die within months of symptom onset. CJD occurs sporadically and belongs to a family of transmissible and universally lethal mammalian diseases.[68] After CJD, Kuru is second-best known within this group of diseases, caused by ritual cannibalism in Papua New Guinea where the Fore tribe used to eat deceased relatives.[69,70] Since cannibalism was stopped in the early 1960s, Kuru lingered due to its long incubation period, but finally declined sharply.[71]

The source of BSE was a new type of cattle feed. The natural cow nutrition of plants was supplemented with meat and bone meals prepared from the offal of cattle, sheep, pigs, and chickens. Such a mixed food had already been in use for some time, but in the late 1970s, a previous production method was replaced to achieve an increased fat content. This modified feed was the source of the spread of prions. In 1994 the first case of vCJD resulting from cattle transmission was identified in man.[64]

In 1996, the UK government announced that a new variant of CJD was probably linked to exposure to the cattle disease BSE, unleashing the most damaging science-based political crisis that has ever occurred in the UK. Ministers and senior officials had insisted for a long time that BSE in British dairy and beef herds posed no threat whatsoever to human health. British beef was alleged "perfectly safe"; regulations went allegedly far beyond what was needed to protect the public; these conclusions were allegedly fully supported by scientific evidence and expertise.[72] Still in 2001, the UK agricultural minister publicly fed his daughter with potentially contaminated meat. When the possible transmission became known, relevant parts of the public lost their faith in the UK governance.[73] Almost immediately after the announcement of the possible transmission of BSE from cattle to humans, the European Union (EU) banned all exports of British beef. The British government introduced new measures to contain BSE, such as a selective cull of cattle reared alongside those with BSE between 1989 and 1993.[74] The global epidemic of BSE-transmitted vCJD peaked in 2000 and then declined dramatically.[66] The number of exposed individuals is especially important for assessing the risk of secondary transmission via blood transfusion, plasma products, or contaminated surgical instruments. The risk of secondary transmission from the asymptotic carriers of prion disease continues.[75] For all these reasons, global continuous observation of

CJD and other more "exotic" types of prion-caused diseases in man and animals (there are many) is important.[66]

12 Intermediate summary

During the 18th and 19th centuries industrial output and urban populations grew rapidly. Authorities balance between citizen complaints and commercial interests. Smoking chimneys and dead rivers were often seen as signs of progress. Technical solutions were tried, such as more efficient boilers, clean coal, or higher chimneys. Some particularly controversial practices like the open-air roasting of ores were banned. Toxic substances not only spread as a result of industrial pollution but also because of public demand. Toxic substances like arsenic, lead, mercury, and new synthetic coal tar dyes entered homes and bodies via wallpapers, paint, cosmetics, and dyed food. Physicians and other healthcare professionals used both vegetable and mineral substances to try to treat various maladies. Investigations into the safety of synthetic additives and sweeteners, inspections for microbial and chemical contamination as well as pasteurization and hygiene requirements followed. In the beginning of the 20th century, today's US Food and Drug Administration (FDA) emerged, although in the beginning under different names. Occupational exposure to toxic substances turned into a further research area, looking not only at acute toxicity but also at low-dose and long-term exposure to substances like lead and new hazards like radiation. After World War II, trust in technical fixes of toxic hazards began to erode. Discussions of toxic exposure had previously been limited to individual sites, practices, and products. Now toxic fears encompassed not only local but also regional and global environments. This shift of concerns was accompanied by demands to protect nature. Scientists and regulators were once again expected to mediate between health concerns and the post-war boom of industrial production, which saw a flood of new substances inundate the global market. While officials frequently established tolerances for "safe" exposure to hazardous substances, cultural taboos, growing cancer fears, and new research on mutagenicity led to demands for zero-tolerance of toxic and carcinogenic substances in food and nuclear testing bans.[41,76] Pesticides such as dichlorodiphenyltrichloroethane (DDT) were banned from use in the US and countries around the world.[77]

Popular environmentalist values became increasingly international. Disasters such as Bhopal and Chernobyl heightened public awareness of global toxic hazards and the long-term costs of contaminated landscapes. Environmental justice movements also highlighted the plight of disadvantaged communities living with toxic burdens.[41]

Despite scientific discussion of dangerous characteristics of many major industrial chemicals, there was little regulation of industry anywhere in the world before the 1970s. The large-scale export of hazardous industries to developing countries coincided with the global expansion of industry in the 1970s. In industrialized countries, new laws began to regulate airborne exposure to toxic substances at work and pollution in the environment. But most new industries set up in developing countries continued with business-as-usual as it had been done for decades in the originating countries, including the export of hazardous and discredited technologies such as the manufacture of asbestos textiles. For a while, corporate double standards existed.[49]

With globalization people, companies, institutions, and governments interact increasingly worldwide. Globalization has accelerated since the last centuries, based on advances in technologies of transport and communication. International trade has grown massively, as has the exchange of ideas, beliefs, and culture. Globalization is both an economic process of interaction and an accelerated exchange of social and cultural aspects. Goods, services, data, technology, and capital are increasingly exchanged worldwide, helped by the reduction of cross-border trade barriers. Multiple layers of infrastructure are necessary for these processes, including advances in transportation from the steam locomotive over steamships to container ships, the telecommunication infrastructure from the telegraph, over the internet to mobile phones and other gadgets.[78]

We read every day about small, medium, or large disasters that happen worldwide or even become direct eyewitnesses on television or the internet. Disasters show us how vulnerable we are when the familiar environment is no longer available as we were used to. They also show us how much we build in our life today on what was painstakingly built up by previous generations and which today provides the basis of our daily life. The COVID-19 pandemic is such a disaster. In contrast to most of the other disasters of which we tend to notice something in the distance, many people around the world are directly exposed to the COVID-19 pandemic.

13 Discussion and conclusions

We can (not yet?) prevent earthquakes, hurricanes, tsunamis, and other major disasters. Disasters range between completely man-made and acts of God. In most cases, human elements prevent prevention and/or proper addressing. The Fukushima nuclear meltdown could have been prevented by a simple precautionary measure, namely the construction of the emergency generator on raised ground, where the tsunami could not have spilled over it.

The COVID-19 disaster hit humanity in a situation where society is already very complex, including elaborate public health systems, advanced healthcare, and advanced science. The individual institutions, such as hospitals, ambulance services, or emergency services, are usually well organized on a technical and local level. But increasingly, we face challenges that we cannot see with the naked eye. Unfiltered perception in photographs, on television, or by eyewitnesses who report on social media usually only recognizes individual aspects and often shows only the tip of the iceberg. The intellectual processing of reality is a social process in which science plays an elementary role. But the disasters discussed above show that in their mental processing science is not as neutral as it likes to pretend to be. For half a century, mainstream science believed that the lead concentrations measured in the environment and blood samples were "natural." But they were hundreds of times higher than in nature before the start of industrialization. They had often reached already acute toxic levels and impaired the health of millions of children and adults. The introduction and maintenance of the Kehoe principle (show me the data) was a masterpiece in the psychological warfare on the level of communication, performed by the car industry and its intellectual representatives.[38,79]

In the described environmental disasters, we can identify three stages. During the first stage, mankind is predominantly unaware of the side effects of new technical means, such as finding out the toxicity of lead which took millennia or smoking, the detrimental properties of which remained hidden for centuries. Once the respective hazards are identified, a second stage begins, with two main players. One side is usually companies that have a material interest in denying the real connections. But the same role was played by farmers who continued to use animal feed containing animal protein that they had bought and stocked earlier, even after feeding it had become illegal and was known to be dangerous. The other side is those persons that was/are directly harmed physically, or who see with their own eyes how their children or their families are harmed. Both sides discuss these things in public. There was and is a high level of trust in science, but smoking and chronic lead poisoning show how science often lags in understanding reality. This can either be due to bribed or otherwise biased scientists, or due to false basic assumptions, such as the belief that there is either acute lead poisoning or none at all. The existence of chronic poisoning by toxic metals, which particularly harm children and adolescents, did not fit into the existing paradigms. It requires a paradigm shift to do justice to reality. In the third phase, legal and administrative measures are taken. Of course, both sides lament continuously that too much or too little is being done. Once affairs have reached the political level, they are subject to the rules of political controversy, including shameless exaggerations, understatements, and other instruments of political theater.

There are different basic attitudes with which one can try to find one's way in our world. One is to trust blindly official statements who assure us that everything is running smoothly. That is the attitude governments prefer. The other option is to keep your eyes and ears open and try to look behind the curtains. A critical attitude towards official whitewashing is one key prerequisite for a worldview based on common sense. A second prerequisite is to know that there are conflicts of interest behind all facades. Institutions' facades usually represent momentary compromises somewhere between the truth and the impression a current majority of decision-makers within the respective institution is trying to make. A third prerequisite is to know that where humans are involved there is no absolute truth. A fourth one is that in our ever more complex world there are irrational movements about everything and everyone, which in times of uncertainty get the opportunity to wider public attention, such as vaccination hesitancy.[80,81] There are more challenges, but we stop here.

Literature has complained of an increasingly complex world since there were written records. The world is indeed getting more and more complex. But children and young people (have to) learn how the world works. Instead of exploring all details in all possible, impossible, and historical directions and dimensions, they take things as they, live with them, and come to terms with them. Sometimes, we realize that assumptions we made decades ago were flawed. Somehow, we need both capabilities: to accept things as they are, like children, and to look critically behind the surface, like seasoned, battle-hardened adults.

Mistakes tend to creep into the interface between institutions and general social trust. These errors can be corrected, but not without both sides, decision-makers and the general public, leaving their respective comfort zones. Some or many prefer not to hear unpleasant truths or to abandon comfortable illusions. Many scientists associate many of today's disasters with global warming, the globalization of the world economy, and the growing size of the world's population. When we look at their warnings, we feel like Armageddon is imminent. A lot of self-interest and pseudo-science are behind these warnings. An early prophet of these warnings was Malthus, who since around 1800 warned of an emerging discrepancy between the growing world population and ability to feed these people and prophesied worldwide catastrophes of disease and famine.[82] He contributed to establishing a critical ideological position towards scientific and technological progress, followed by many more alarming warnings, including those of the Club of Rome,[83] and of today's Greta Thunberg, who demands world leaders should take immediate action against global warming and climate change.[84] But reality is more complex than a pubertal adolescent wants to believe.

An additional aspect of disasters is that knowledge of harmfulness alone does not prevent everyone from changing harmful habits, as we see with smoking.

References

1. Janis IL. *Groupthink: Psychological Studies of Policy Decisions and Fiascoes.* Boston, MA: Houghton Mifflin; 1982.
2. Janis IL. *Victims of Groupthink: A Psychological Study of Foreign Policy Decisions and Fiascoes.* Boston, MA: Houghton Mifflin; 1972.
3. Wikipedia. *Groupthink.* https://en.wikipedia.org/wiki/Groupthink.
4. Hart P, Janis IL. Victims of groupthink. *Polit Psychol.* 1991;12(2):247–278. https://www.researchgate.net/publication/273109291_Irving_L_Janis'_Victims_of_Groupthink/link/5c51fbc192851c22a39bdcce/download.
5. Wikipedia. *Attack on Pearl Harbour.* https://en.wikipedia.org/wiki/Attack_on_Pearl_Harbor.
6. https://www.britannica.com/event/Pearl-Harbor-attack/The-attack.
7. Wikipedia. *Bay of Pgs Invasion.* https://en.wikipedia.org/wiki/Bay_of_Pigs_Invasion.
8. Wikipedia. *Space Shuttle Challenger Disaster.* https://en.wikipedia.org/wiki/Space_Shuttle_Challenger_disaster.
9. Price MR, Williams TC. When doing wrong feels so right: normalization of deviance. *J Patient Saf.* 2018;14(1):1–2.
10. Vaughan D. *The Challenger Launch Decision: Risky Technology, Culture, and Deviance at NASA.* Chicago, IL: University of Chicago Press; 1996:1–7. 190–195.
11. The Cost of Silence: Normalization of Deviance and Groupthink. *Senior Management ViTS Meeting November 3, 2014.* Terry Wilcutt Chief, Safety and Mission Assurance HalBell Deputy Chief, Safety and Mission Assurance; 2014. https://sma.nasa.gov/docs/default-source/safety-messages/safetymessage-normalizationofdeviance-2014-11-03b.pdf?sfvrsn=4.
12. *Symptoms of Group Think.* http://www.geocities.ws/oralcompgroupthink/symptoms.htm.
13. Wikipedia. *Space Shuttle Columbia Disaster.* https://en.wikipedia.org/wiki/Space_Shuttle_Columbia_disaster.
14. Woodyard C. *FAA Certifies Troubled Boeing 737 Max to Fly Again, But Critics Fear Jetliner's 'Basic Aerodynamic Problem'.* USA Today; 2021.
15. Levin A, Ryan C, Philip S. Last to ground the max a year ago, FAA's global status shaken. *Claims J.* 2020;(March 18). https://www.claimsjournal.com/news/international/2020/03/18/296074.htm.
16. Wikipedia. *Boeing 737 MAX Groundings.* https://en.wikipedia.org/wiki/Boeing_737_MAX_groundings.
17. Gates D. *Boeing's 737 MAX 'Design Failures' and FAA's 'Grossly Insufficient' Review Slammed*; 2020. https://www.seattletimes.com/business/boeing-aerospace/u-s-house-preliminary-report-faults-boeing-faa-over-737-max-crashes/.
18. Lord of Liverpool Russell. *The Knights of the Bushido (A Short History of Japanese War Crimes).* London, UK: Cassell & Co; 1958.
19. Wikipedia. *2011 Tōhoku Earthquake and Tsunami.* https://en.wikipedia.org/wiki/2011_T%C5%8Dhoku_earthquake_and_tsunami.
20. Wikipedia. *Fukushima Daiichi Nuclear Deaster.* https://en.wikipedia.org/wiki/Fukushima_Daiichi_nuclear_disaster.
21. Needleman H.L. History of Lead Poisoning in the World. n.d. https://www.biologicaldiversity.org/campaigns/get_the_lead_out/pdfs/health/Needleman_1999.pdf.
22. Hanna-Attisha M, Lanphear B, Landrigan P. Lead poisoning in the 21st century: the silent epidemic continues. *Am J Public Health.* 2018;108(11):1430. https://www.ncbi.nlm.nih.gov/pmc/articles/PMC6187797/pdf/AJPH.2018.304725.pdf.
23. Pearce JM. Burton's line in lead poisoning. *Eur Neurol.* 2007;57(2):118–119.
24. Hernberg S. Lead poisoning in a historical perspective. *Am J Ind Med.* 2000;38(3):244–254.

25. E. Sohn. Lead: Versatile Metal, Long Legacy n.d. https://sites.dartmouth.edu/toxmetal/more-metals/lead-versatile-metal-long-legacy/.
26. Riva MA, Lafranconi A, D'Orso MI, et al. Lead poisoning: historical aspects of a paradigmatic "occupational and environmental disease". *Saf Health Work*. 2012;3(1):11–16. https://www.ncbi.nlm.nih.gov/pmc/articles/PMC3430923/pdf/shaw-3-11.pdf.
27. Tisma M. "Rich man, poor man": a history of lead poisoning. *Hektoen Int J*. 2019. https://hekint.org/2019/09/24/rich-man-poor-man-a-history-of-lead-poisoning/.
28. Miracle VA. Lead poisoning in children and adults. *Dimens Crit Care Nurs*. 2017;36(1):71–73.
29. Needleman HL. The removal of lead from gasoline: historical and personal reflections. *Environ Res*. 2000;84(1):20–35.
30. Nriagu JO. Saturnine gout among Roman aristocrats. Did lead poisoning contribute to the fall of the empire? *N Engl J Med*. 1983;308(11):660–663.
31. Weitzman M. American pediatric society's 2017 John Howland award acceptance lecture: a tale of two toxicants: childhood exposure to lead and tobacco. *Pediatr Res*. 2018;83(1–1):23–30.
32. Zak L. *A Brief Dive Into the History of Lead Paint*. Zota Pro; July 10, 2020. https://zotapro.com/blog/lead-paint-history/.
33. Wikipedia. *Clair Cameron Patterson*. https://en.wikipedia.org/wiki/Clair_Cameron_Patterson.
34. Wikipedia. *Herbert Needleman*. https://en.wikipedia.org/wiki/Herbert_Needleman.
35. Kovarik W. Ethyl-leaded gasoline: how a classic occupational disease became an international public health disaster. *Int J Occup Environ Health*. 2005;11(4):384–397.
36. C. Nickson Precautionary Principle and the Kehoe Principle n.d. https://litfl.com/precautionary-principle-and-the-kehoe-principle/.
37. Wikipedia. *Robert A Kehoe*. https://en.wikipedia.org/wiki/Robert_A._Kehoe.
38. Nriagu JO. Clair Patterson and Robert Kehoe's paradigm of "show me the data" on environmental lead poisoning. *Environ Res*. 1998;78(2):71–78.
39. Hagner C. *Historical Review of European Gasoline Lead Content Regulations and Their Impact on German Industrial Market*. GKSS Forschungszentrum Geesthacht; 1999. https://www.osti.gov/etdeweb/servlets/purl/20018469.
40. Wikipedia. *Love Canal*. https://en.wikipedia.org/wiki/Love_Canal.
41. Kirchhelle C. Toxic tales—recent histories of pollution, poisoning, and pesticides (ca. 1800–2010). *NTM*. 2018;26:213–239. https://www.ncbi.nlm.nih.gov/pmc/articles/PMC5993852/pdf/48_2018_Article_190.pdf.
42. Jackson AC. Chronic neurological disease due to methylmercury poisoning. *Can J Neurol Sci*. 2018;45(6):620–623.
43. Ruggieri F, Majorani C, Domanico F, et al. Mercury in children: current state on exposure through human biomonitoring studies. *Int J Environ Res Public Health*. 2017;14(5):519. https://www.ncbi.nlm.nih.gov/pmc/articles/PMC5451970/pdf/ijerph-14-00519.pdf.
44. Watanabe C, Satoh H. Evolution of our understanding of methylmercury as a health threat. *Environ Health Perspect*. 1996;104(Suppl. 2):367–379. https://www.ncbi.nlm.nih.gov/pmc/articles/PMC1469590/pdf/envhper00345-0195.pdf.
45. Harada M. Minamata disease: methylmercury poisoning in Japan caused by environmental pollution. *Crit Rev Toxicol*. 1995;25(1):1–24.
46. Wikipedia. *Minamata Disease*. https://en.wikipedia.org/wiki/Minamata_disease.
47. *Minamata Convention on Mercury*. https://www.mercuryconvention.org/.
48. Wikipedia. *Minamata Convention on Mercury*. https://en.wikipedia.org/wiki/Minamata_Convention_on_Mercury.
49. Castleman B. The export of hazardous industries in 2015. *Environ Health*. 2016;15:8. https://www.ncbi.nlm.nih.gov/pmc/articles/PMC4717658/pdf/12940_2016_Article_91.pdf.
50. Castaldelli-Maia JM, Ventriglio A, Bhugra D. Tobacco smoking: from 'glamour' to 'stigma'. A comprehensive review. *Psychiatry Clin Neurosci*. 2016;70(1):24–33. https://onlinelibrary.wiley.com/doi/epdf/10.1111/pcn.12365.

References

51. Tobacco Industry Research Committee (TIRC). *Tobacco Tactics*; February 7, 2020. https://tobaccotactics.org/wiki/tobacco-industry-research-committee/.
52. Council for Tobacco Research. SourceWatch; December 25, 2019. https://www.sourcewatch.org/index.php/Council_for_Tobacco_Research.
53. Wikipedia. *Tobacco Institute.* https://en.wikipedia.org/wiki/Tobacco_Institute.
54. Wikipedia. *Center for Indoor Air Research (CIAR).* https://en.wikipedia.org/wiki/Center_for_Indoor_Air_Research.
55. Wikipedia. *Tobacco Master Settlement Agreement.* https://en.wikipedia.org/wiki/Tobacco_Master_Settlement_Agreement.
56. Niemeyer D, Miner KR, Carlson LM, et al. The 1998 Master Settlement Agreement: a public health opportunity realized—or lost? *Health Promot Pract.* 2004;5(3 Suppl):21S–32S.
57. Cole HM, Fiore MC. The war against tobacco: 50 years and counting. *JAMA.* 2014;311(2):131–132. https://www.ncbi.nlm.nih.gov/pmc/articles/PMC4465196/pdf/nihms696825.pdf.
58. CDC. *History of the Surgeon General's Reports on Smoking and Health;* 2019. https://www.cdc.gov/tobacco/data_statistics/sgr/history/index.htm.
59. Brawley OW, Glynn TJ, Kuri FR, et al. The first surgeon general's report on smoking and health: the 50th anniversary. *CA Cancer J Clin.* 2014;64(1):5–8. https://acsjournals.onlinelibrary.wiley.com/doi/epdf/10.3322/caac.21210.
60. Viscusi WK, Hersch J. In: Kessler DP, ed. *Tobacco Regulation Through Litigation: The Master Settlement Agreement. Regulation vs. Litigation: Perspectives From Economics and Law.* Chicago, ILL, USA: University of Chicago Press; 2010. https://www.nber.org/system/files/chapters/c11959/c11959.pdf.
61. Clark TT, Sparks MJ, McDonald TM, et al. Post-tobacco master settlement agreement: policy and practice implications for social workers. *Health Soc Work.* 2011;36(3):217–224.
62. Blum A, Wolinski H. AMA rewrites tobacco history. *Lancet.* 1995;346(8970):261.
63. Editorial. Tobacco and the A.M.A. *N Engl J Med.* 1964;270:959–960. https://doi.org/10.1056/NEJM196404302701811.
64. Norrby E. Prions and protein-folding diseases. *J Intern Med.* 2011;270(1):1–14.
65. Wikipedia. *Prion.* https://en.wikipedia.org/wiki/Prion.
66. Watson N, Brandel JP, Green A, et al. The importance of ongoing international surveillance for Creutzfeldt-Jakob disease. *Nat Rev Neurol.* 2021;17(6):362–379.
67. Wikipedia. *Variant Creutzfeldt–Jakob Disease.* https://en.wikipedia.org/wiki/Variant_Creutzfeldt%E2%80%93Jakob_disease.
68. Pearce JM. Jakob-Creutzfeldt disease. *Eur Neurol.* 2004;52(3):129–131. https://www.karger.com/Article/Pdf/81462.
69. Liberski PP. Historical overview of prion diseases: a view from afar. *Folia Neuropathol.* 2012;50(1):1–12. https://www.termedia.pl/Review-paper-r-n-r-nHistorical-overview-of-prion-diseases-a-view-from-afar,20,18385,1,1.html.
70. Liberski PP. Kuru: a journey back in time from Papua New Guinea to the neanderthals' extinction. *Pathogens.* 2013;2(3):472–505. https://www.ncbi.nlm.nih.gov/pmc/articles/PMC4235695/pdf/pathogens-02-00472.pdf.
71. Mahat S, Asuncion RMD, Kuru. *StatPearls [Internet].* Treasure Island, FL: StatPearls Publishing; January 16, 2021. https://www.ncbi.nlm.nih.gov/books/NBK559103/.
72. Millstone E, van Zwanenberg P. Politics of expert advice: lessons from the early history of the BSE saga. *Sci Public Policy.* 2001;28(2):99–112. http://citeseerx.ist.psu.edu/viewdoc/download?doi=10.1.1.943.1268&rep=rep1&type=pdf.
73. Martin C. UK Government's handling of mad cow disease. *Lancet Neurol.* 2015;14(8):793.
74. Ainsworth C. Damian Carrington BSE disaster: the history. *New Sci.* 2000;(25 October). https://www.newscientist.com/article/dn91-bse-disaster-the-history/.
75. Chen CC, Wang YH. Estimation of the exposure of the UK population to the bovine spongiform encephalopathy agent through dietary intake during the period 1980 to 1996. *PLoS One.* 2014;9(4), e94020. https://www.ncbi.nlm.nih.gov/pmc/articles/PMC3988046/pdf/pone.0094020.pdf.

76. Carson R. *Silent Spring*. Boston, MD: Houghton Mifflin; 1962.
77. Wikipedia. *DDT*. https://en.wikipedia.org/wiki/DDT.
78. Wikipedia. *Globalization*. https://en.wikipedia.org/wiki/Globalization.
79. Berg J. *A Grim Very Tale: The Kehoe Paradigm*. Countercurrents; November 6, 2015. https://www.countercurrents.org/berg060115.htm.
80. Wikipedia. *Vaccination Hesitancy*. https://en.wikipedia.org/wiki/Vaccine_hesitancy.
81. Di Pietro ML, Posia A, Teleman AA, et al. Vaccine hesitancy: parental, professional and public responsibility. *Ann Ist Super Sanita*. 2017;53(2):157–162. https://www.iss.it/documents/20126/45616/ANN_17_02_13.pdf.
82. Wikipedia. *Thomas Robert Malthus*. https://en.wikipedia.org/wiki/Thomas_Robert_Malthus.
83. Wikipedia. *Club of Rome*. https://en.wikipedia.org/wiki/Club_of_Rome.
84. Wikipedia. *Greta Thunberg*. https://en.wikipedia.org/wiki/Greta_Thunberg.

CHAPTER

10

Basic research, applied research, and the real world

Deciphering the human genome was a huge project that officially ran from 1990 to 2003. It has certainly advanced human understanding of the rules according to which cells multiply, how genetic information is passed on to the next generation, and how the experiences and conflicts of all generations of humans and their precursors in nature are encoded in our genes. As this project unfolded, also many misconceptions had to be corrected, including a mechanistic mindset that believed that deciphering the human genome could answer the essential questions of human life. On closer inspection, however, this project also revealed a side of science and research that scientists and researchers are less likely to talk about. In their working world, scientists are in a different world, but they also need a stable work structure, social recognition, and the prospect of permanent employment. They share the latter side with all other people. One consequence is that where science is mainly financed from public sources, unwritten rules creep into the administrative side of science and research. Other factors are prejudices, arrogance, and more. All these factors together may increase the comfort of work in the short term, but they also tend to increase the bureaucratic and administrative side of science and research and to reduce competitiveness. We see today how women, members of social minorities, and other minorities have been discriminated against in the past. These things have certainly improved, but we can also look at the history of the human genome project as a textbook example of how it was necessary to disrupt the calm of the scientific community to allow new, better methods to enter research.

The Human Genome Project was an international scientific research project to determine the human DNA and to map the genes of the human genome. The idea of sequencing the entire human genome was first proposed in discussions in the US from 1984 to 1986. A committee appointed by the US National Research Council endorsed the concept, but recommended a broader program to include the creation of genetic, physical, and sequence maps of the human genome; efforts in key model organisms including bacteria, worms, flies, and mice; to develop technology for these objectives; and research into the ethical, legal and social challenges of such a project. The Human Genome Project was launched in the US as a joint effort of the Department of Energy and the National Institutes of Health (NIH). Further support came from institutions in the UK, France, Japan, the EU, and later Germany and China.[1]

It was so far the world's largest collaborative biological project. Launched in 1990, it was declared complete in 2003. Most of the government-sponsored sequencing was performed in 20 universities and research centers in the USA, the UK, Japan, France, Germany, and China. It originally aimed to map the nucleotides in a human reference genome (more than 3 billion). The genome of any given individual is unique; mapping the "human genome" involved sequencing some individuals and then assembling to get a complete sequence for each chromosome. The finalized human genome is a mosaic that does not represent any one individual. However, the vast majority of the human genome is the same in all humans. The idea for obtaining a reference sequence had three independent origins in the world of science. It became reality when US government funding was obtained through political support. When the $3 billion projects were formally founded in 1990, it was expected to take 15 years. But the first round of essentially complete genome was published already in 2003, 2 years earlier than planned. Other additional milestones followed. Key findings were that there are a bit more than 22,000 protein-coding genes in humans, comparable to other mammals, and much less than thought originally; the human genome has also many more duplications (nearly identical, repeated sections of DNA) than had been previously thought; fewer than 7% of protein families appeared to be vertebrate-specific.[2,3]

In the 1980s, mapping of important disease genes had been achieved, specifically the location of three important disease genes: those responsible for cystic fibrosis, Duchenne muscular dystrophy, and Huntington's disease. By the late 1980s, multiple approaches for sequencing DNA were in use, but costs and time constraints were still a limiting factors. However, this all began to change with the work of the NIH scientist John Craig Venter. For several years, he had been using automated DNA sequencers to sequence portions of chromosomes associated with specific diseases. In 1991, he described how he used his high-tech equipment to sequence more than 600 expressed sequence tags (ESTs) from a brain complementary DNA collection, identifying about half of them as genes, far more than anyone else had ever reported in a single paper to date.[4] In this approach, DNA is partially sequenced, or "tagged," using an automated DNA sequencing machine. The resulting sequences are long enough for one to be distinguished from another.[5]

His 1991 paper make an impact, as did his claims that in his laboratory alone he could sequence as many as 10,000 ESTs a year at a low cost.[4] In 1992, he published the sequences of more than 2000 genes, which brought the total to 2500 genes sequenced in one laboratory, which was as many as had been sequenced in the entire world to that point.[6,7]

Venter had been drafted as a Navy Corpsman in Vietnam from 1967 to 1968, where he worked in the intensive-care ward of a field hospital. Being confronted with severely injured and dying marines daily made him desire to study medicine, but he later switched to biomedical research. After studying, he joined the National Institutes of Health (NIH). Venter was passionate about the power of genomics to radically transform healthcare. Venter believed that shotgun sequencing was the fastest and most effective way to get useful human genome data.[8-10] In shotgun sequencing, DNA is broken up randomly into numerous small segments, which are sequenced using the chain termination method to obtain *reads*. Multiple overlapping reads for the target DNA are obtained by performing several rounds of this fragmentation and sequencing. Computer programs then use the overlapping ends of different reads to assemble them into a continuous sequence. Shotgun sequencing was one of the precursor technologies that was responsible for enabling whole genome sequencing.[11] Such shotgun sequencing technique was rejected by the Human Genome Project, however, as several scientists thought it would not be accurate enough for a genome as complicated as that of humans, that it would be logistically more difficult, and that it would cost significantly more.[2,8,9]

Venter had seen the slow pace of progress in the Human Genome Project as an opportunity to continue his interest in patenting genes, so he sought funding from the private sector to start Celera Genomics. In 1992, he left the NIH and started with the help of venture capital a nonprofit research institute at which he quickly established several automated sequencing machines. With his newly founded "The Institute for Genomic Research" (TIGR, now part of the J Craig Venter Institute),[9] he decided to sequence an entire organism, which was much more complex than any yet attempted. For this, he used "whole-genome random sequencing." *Haemophilus influenzae* ("influenza" is usually abbreviated as "flu") is a bacterium that can cause ear and respiratory infections. The size of its genome was fairly typical for a bacterium, but about 10 times longer than the genome of any virus that had been sequenced so far. "Whole-genome random sequencing" was a stepwise process that aimed to assemble a wholly sequenced genome from partly sequenced DNA fragments with the help of a computational model. With this approach, a preliminary physical map of the genome was no longer needed. Copies of DNA from *H. influenzae* were cut into pieces, were then partly sequenced at both ends, using automated sequencing machines, revealing "read lengths" each several 100 base pairs. These base-pair sequences—with their many overlaps—became the raw data that was entered into the computer. Smaller libraries of longer fragments—15,000–20,000 base pairs—were also developed. Using the "TIGR assembler," a software tool, the many thousands of fragments were compared, clustered, and matched for assembling the genome. The most informative and nonrepeating sequences were identified first. Repeating fragments were compared next. The longer fragments helped to order some of the very repetitive and almost identical sequences. Small physical gaps that remained after the TIGR assembler performed its work were rectified with additional auxiliary strategies. Assembling the *H. influenzae* genome from over 20,000 DNA fragments was a considerable achievement, and surprised many. Once assembled, the genetic coding regions were located, compared to known genes, and a detailed map was developed. Fleischmann, Venter, and others published their results in the journal *Science* in 1995.[12,13]

In 1996, Venter published a paper, "A new strategy for genome sequencing," arguing that the advances made in expressed sequence tags (EST) and mapping a bacterial genome made it feasible to apply the method to the entire human genome and to complete the project before 2005.[14,15]

Noting that critics opposed the EST method of genome sequencing, in 1998 Venter founded Celera Genomics, a for-profit company. Venter planned to sequence the human genome using tools and techniques that he and his team developed. Celera worked with TIGR, which focused on decoding and analyzing the genomes of several organisms, including the bacteria responsible for Lyme disease, syphilis, tuberculosis, and meningitis, while Celera focused on the human genome. Venter's approach differed from that of the NIH. With these two approaches running parallel to one another, the US Congress called Francis Collins, the director of the National Human Genome Research Institute (NHGRI) of the NIH, and Venter, of Celera, to testify before Congress in Washington D.C. in 1998. With Venter's efforts underway, Congress asked why it should continue to fund the NIH genome efforts. Venter defended his approach and told Congress that his company did not intend to patent most of their data, except a few 100 sequences that were sufficiently new, not obvious, and useful. Congress had no problem with Venter's approach. After the hearings, the race between Celera and the NIH to map the human genome accelerated.[15]

In the end, Venter and his team at Celera Corporation shared credit for sequencing the first draft human genome with the publicly funded Human Genome Project. A 'rough draft' of the genome was finished in 2000 and was announced jointly by US President Bill Clinton and British Prime Minister Tony Blair in June 2000.[2,8] In 2001, the Human Genome Project published the first Human Genome in the journal *Nature*,[1] followed 1 day later by a Celera publication in the journal *Science*.[16] It must have been an amazing feeling on the part of the scientists at the Human

Genome Project as their venerable position was challenged by a newcomer and rebel, which probably many of them perceived as a daredevil.

Despite some claims that shotgun sequencing was less accurate than the clone-by-clone method chosen by the Human Genome Project, the technique became widely accepted by the scientific community.[8] In his book "Life at the Speed of Light" Venter claimed later a convergence between the code of DNA and digital computer code, which would eventually result in a coming age of digital biology, in which a genome could be designed on a computer, transmitted electronically, and reconstituted in a rapid synthesizer.[17] The availability of these techniques has played a key role in the publication of the genome of the SARS-CoV-2 virus.

Several additional follow-up reports about "more" finalizing of the Human Genome Project have been published since 2003.[18–21]

Progress in genomics in the last decade has brought new methods for massively parallel sequencing of large numbers of short DNA sequences. These methods, often referred to as next-generation sequencing, are used for genome sequencing, but their high throughput also makes them very attractive for functional genomics investigations, where the primary goal is to link genome regions with their biological function.[22]

There are now many new projects that follow the Human Genome Project, which involve from a few tens to several million genomes currently in progress, which vary from having specialized goals or a more general approach.[23]

The history of the Human Genome Project is important to understanding science in today's complex world. It also helps us better appreciate the value of initiative and entrepreneurship. Venter would not have survived long in an authoritarian country like Russia or China. He would have been fired, would either have had a "tragic" accident, would have simply disappeared, would have been charged with some fabricated crime and thrown in jail, or would have been disposed of in some other way. The second essential element is the lack of respect of venture capital for venerable authorities. If there is potentially money to be made with a good idea, venture capital can be found, no matter how many terribly important professors (TIPs) yell against it. It is not the first time in the history of science that a good idea has been rejected by the vast majority of the well-established mainstream.

The Volkswagen (VW) emissions scandal began when the US Environmental Protection Agency (EPA) issued a notice of violation of the US Clean Air Act. The EPA had found that VW had intentionally programmed its diesel engines to activate their emissions controls only during laboratory testing, which caused the vehicles' emissions to be OK during testing. But the emission during real-world driving was 40 times above the allowed limit.[24] The true scandal behind this story is that in Germany, which otherwise has a strong focus on environmental protection, neither the supervisory authorities intervened in time, nor did the law enforcement authorities intervene immediately.[25] Regulations are only as good as they are enforced. There is probably no other country on earth in which as much research is carried out on environmental protection as in Germany. And yet it was the US who brought the Diesel software fraud scandal to light, and not Germany. CEOs of large companies enjoy similarly high esteem in Europe as university professors. They can do things that normal mortals cannot do. This leads us to the question of how seriously the assurances of the German authorities regarding environmental protection, honesty, and sincerity are to be taken. But this is another story.

References

1. International Human Genome Sequencing Consortium. Initial sequencing and analysis of the human genome. *Nature*. 2001;409:860–921. https://www.nature.com/articles/35057062.pdf.
2. Wikipedia. *Human Genome Project*. https://en.wikipedia.org/wiki/Human_Genome_Project.
3. Moraes F, Góes A. A decade of human genome project conclusion: scientific diffusion about our genome knowledge. *Biochem Mol Biol Educ*. 2016;44(3):215–223. https://iubmb.onlinelibrary.wiley.com/doi/epdf/10.1002/bmb.20952.
4. Adams MD, Kelley JM, Gocayne JD, et al. Complementary DNA sequencing: expressed sequence tags and human genome project. *Science*. 1991;252(5013):1651–1656.
5. Shampo MA, Kyle RA. J. Craig Venter—the human genome project. *Mayo Clin Proc*. 2011;86(4):e26–e27. https://www.ncbi.nlm.nih.gov/pmc/articles/PMC3068906/pdf/mayoclinproc_86_4_023.pdf.
6. Adams MD, Dubnick M, Kerlavage AR, et al. Sequence identification of 2,375 human brain genes. *Nature*. 1992;355(6361):632–634.
7. Adams J. Sequencing human genome: the contributions of Francis Collins and Craig Venter. *Nat Educ*. 2008;1(1):133. https://www.nature.com/scitable/topicpage/sequencing-human-genome-the-contributions-of-francis-686/.
8. Wikipedia. *Craig Venter*. https://en.wikipedia.org/wiki/Craig_Venter.
9. J Craig Venter Institute. https://www.jcvi.org.
10. J Craig Venter. *Biography*. https://www.jcvi.org/about/j-craig-venter.
11. Wikipedia. *Shotgun Sequencing*. https://en.wikipedia.org/wiki/Shotgun_sequencing.
12. Fleishmann RD, Adams MD, White O, et al. Whole-genome random sequencing and assembly of Haemophilus influenzae Rd. *Science*. 1995;269:496–512. https://www.researchgate.net/publication/15655914_Whole-Genome_Random_Sequencing_and_Assembly_of_Haemophilus_Influenzae_Rd.
13. *Genome News Network (GNN) Genetics and Genomics Timeline*; 1995. http://www.genomenewsnetwork.org/resources/timeline/1995_Haemophilus.php.
14. Venter JC, Smith HO, Hood L. A new strategy for genome sequencing. *Nature*. 1996;381(6581):364–366.
15. Carvalho T. *John Craig Venter. The Embryo Project Encyclopedia*; May 6, 2014. https://embryo.asu.edu/pages/john-craig-venter-1946.
16. Venter JC, Adams MD, Meyers EM, et al. The sequence of the human genome. *Science*. 2001;291(5507):1304–1351. https://science.sciencemag.org/content/291/5507/1304/tab-pdf.
17. Craig Venter J. *Life at the Speed of Light. From the Double Helix to the Dawn of Digital Life*. New York, USA: Viking Books; 2013.

18. Pearson H. Human genome completed (again). *Nature.* 17 May 2006. https://www.nature.com/news/2006/060515/full/news060515-12.html.
19. Nurk S., Koren S., Rhie A., et al. The Complete Sequence of a Human Genome. Biorxiv, the Preprint Server for Biology. n.d. https://www.biorxiv.org/content/10.1101/2021.05.26.445798v1.
20. Reardon S. *A Complete Human Genome Sequence Is Close: How Scientists Filled in the Gaps. Researchers Added 200 Million DNA Base Pairs and 115 Protein-Coding Genes—But They've Yet to Entirely Sequence the Y Chromosome.* Nature News; June 4, 2021. https://www.nature.com/articles/d41586-021-01506-w.
21. LeMieux J. *The Human Genome Project in 2020 Hindsight-Two Decades After the Working Draft of the Human Genome Was Completed, It Is Clearer Than Ever That Analysis of the Text Is Just Getting Started.* Genetic Engineering and Biotechnology News; June 5, 2020. https://www.genengnews.com/insights/the-human-genome-project-in-2020-hindsight/.
22. Guigo R, de Hoon M. Recent advances in functional genome analysis. *F1000Res.* 2018;7. F1000 Faculty Rev-1968 https://www.ncbi.nlm.nih.gov/pmc/articles/PMC6305211/pdf/f1000research-7-16639.pdf.
23. Carrasco-Ramiro F, Peiró-Pastor R, Aguado B. Human genomics projects and precision medicine. *Gene Ther.* 2017;24(9):551–561. https://www.researchgate.net/publication/319115056_Human_genomics_projects_and_precision_medicine/link/59f0b83caca272cdc7cdfb2e/download.
24. Phys.org. *Five things to know about VW's 'dieselgate' scandal;* 2018. https://phys.org/news/2018-06-vw-dieselgate-scandal.html.
25. Wikipedia. *Volkswagen Emissions Scandal.* https://en.wikipedia.org/wiki/Volkswagen_emissions_scandal.

CHAPTER 11

Conflicts of interest and the self-picture of medicine and scientists

1 Introduction

Science (Latin Scientia, knowledge) is the endeavor to systematically create, document, and advance knowledge about our world. It has become a systematic business. Some definitions focus on the physical world only. The Encyclopedia Britannica sees science as the "pursuit of knowledge covering general truths or the operations of fundamental laws."[1] Others divide science into several branches, such as the natural sciences (biology, chemistry, medicine), social sciences (economy, psychology, sociology), and formal sciences (mathematics, computer science).[2] Comparable but slightly different categorizations are on the websites of leading scientific journals.[3–7] Science is based on research, commonly conducted by scientists in academic and research institutions, government agencies, and companies.[2] Reproducibility and replicability of the scientific results are core concepts that allow others to check and reproduce the results under the same or at least similar conditions described in a publication and produce similar results with the same or similar measurements.[8]

Traditionally, science was seen as continuously growing, expanding steadily the extent of human knowledge. This static view was challenged by Thomas Kuhn who began to break up science into three stages, beginning with prescience, without a central paradigm; followed by "normal science," which for a while is quite productive. Over time, science then reaches a crisis, at which point a new paradigm emerges. He termed this revolutionary science and introduced the concept of paradigm shifts in the development of science.[9] The most used example for such a shift was the transition from seeing the earth as the center of the entire world toward a view which placed the sun at the center of our planetary system and turned the medieval worldview upside down. However, there have been several further paradigm shifts in the 20th century, such as the "scientifically proven" belief that lead in car gasoline is harmless. It took about half a century until the Kehoe principle that absence of evidence of risk is evidence of the absence of risk was refuted.[10] Robert Kehoe had been hired for General Motors to examine health issues related to the production of tetraethyl lead (TEL). The anti-knocking property of TEL as an additive to gasoline allowed cars with higher compression ratios. In the early days of TEL production, workers at the Ethyl plants had fallen ill and a number of them died from lead poisoning. Kehoe was hired to develop protocols for the workers handling TEL. Eventually, he became the chief medical advisor of the Ethyl Corporation and held this position until his retirement. He became a leader in occupational health and was the foremost medical advocate for the use of TEL as an additive in gasoline. Two key researchers eventually contributed to the decline of the Kehoe principle: Clair Cameron Patterson showed that the contemporary concentration of lead in the environment was not natural, but reflected decades of lead release from car gasoline.[11,12] Herbert Needleman showed that the level of lead in fallen out milk teeth of US schoolchildren reflected the lead exposure and that the level of lead exposure correlated negatively with their school performance and hence with their later chances to get good jobs.[13–18] The defense of lead admixture in gasoline and the "scientific" fight against the growing understanding that smoking promotes cancer and other serious diseases are just two examples of how mainstream science was massively influenced for several decades by the harmful interests of the auto- and tobacco industry. It had been smart moves from these industries to build up key scientists to defend against growing knowledge.

Natural sciences such as physics, astronomy, or chemistry explore processes in the inanimate nature. In contrast, in medicine and life sciences contemporary values play a major role. A few decades ago, homosexuality was still considered a disease in the Diagnostic and Statistical Manual of Mental Disorders (DSM) of the American Psychiatric Association (APA).[19–21] The DSM is used for the classification of mental disorders. It is used mainly in the USA, but also internationally by clinicians, researchers, regulatory authorities, health insurance companies, pharmaceutical companies, the legal system, and policymakers.[19] In many countries, homosexuality is still officially banned and those who cannot afford to leave the country or do not want to leave have to live out their sexuality in secret. Erectile dysfunction would have been an unimaginable medical diagnosis half a century ago. Today the drugs to treat it represent a considerable market. Just mentioning female sexual dysfunction would have been a taboo a few decades ago. Today you can talk about it openly. Social sciences are shaped even further and deeper by our social, political, and personal attitudes.

All scientists need to be perceived as scientists and not just as people who express an opinion. The number of scientific publications shows a current state of explosion. PubMed, the free search engine run by the US National Library of Medicine (NLM) that lists references and abstracts on life sciences and biomedical topics, returned upon the search term "science" almost 6 million hits in October 2021.[22] A recent analysis classified the growth rate of scientific publications since the 16th century into three phases, each leading to a tripling of growth rate compared to the previous phase: from a growth rate of less than 1% up to the 18th century to 2%–3% between the world wars, and to 8%–9% from after world war II to 2012.[23] Another study uses as indicators (1) the number of doctoral degrees worldwide, (2) the number of patents issued worldwide, and (3) the number of articles published worldwide. It confirms the exponential growth of science, concludes that science and technology have drastically transformed our lives, states that never before so many people worked intending to better understand how the world works, and concludes that it is difficult to grasp the future pace of innovation.[24,25]

But looking just at numbers explains little. Several more dimensions need to be considered. Institutionalization began in the scholarly communities of the academies, the first ones in Italy, in contrast to religious monasteries and other political, religious, or philosophical institutions.[26,27] From the 19th century on, science became also professionalized. In the Royal Society, the number of academic scientists more than doubled between 1881 and 1914, when they made up 61 percent of the total, while other categories such as "distinguished laymen," soldiers, and clergy were drastically reduced. The process of professionalization not only led to the recognition of status but also to an income as a scientist. This social legitimation occurred earlier in the USA than in Europe. After World War II, the function of scientists devoting themselves full time to research became fully recognized in most western industrialized countries. Research programs became more expensive and more ambitious.[27] The Manhattan Project that developed the first atomic bombs marked an irreversible turning-point in the relations between science and the state: the establishment of science as a national asset, the intervention of governments in the direction and range of research, and the recruiting of researchers for large-scale programs. There were 100,000 researchers (scientists, engineers, and technicians) in the world in 1940, but 10 times this number 20 years later. The Organization for Economic Co-operation and Development (OECD) is an intergovernmental economic organization with 38 member countries, including the US, Canada, Australia, and the major European countries including Switzerland. OECD members are high-income economies and are regarded as developed countries (of course, you can also be poor in an OECD country).[28] In the OECD area alone, the total R&D personnel was estimated at 1,754,430 in 1983, of which the US accounted for a little more than 700,000.[27] Ninety percent of all the scientists that ever lived are alive today.[24]

In the past, rulers were attributed physical healing powers and were asked to heal wounds and ailments by laying on their hands. Today we would call such an attitude superstition. Everything that is certified by scientific research enjoys today a higher authority. Science enjoys a status that sets it apart from the private expression of opinion. This leads to yet another dimension that plays a key role in the context of drugs and vaccines: the certification of clinical studies by the regulatory authorities, which then turns into approval.

All people with an academic degree that work in science or a related area are proud of their status, and usually, they also have a good salary. Universities in the USA, Canada, and Europe are different institutions, but their degrees and are broadly accepted worldwide. While in the USA the position of a university professor is nothing special, most European countries ascribe almost God-like characteristics to such a position. The administrative division of mankind into adult and "pediatric" populations, defined by being before or after the 18th birthday, is currently accepted by academic research, regulatory agencies, and virtually everyone working in healthcare.

Science goes through development cycles.[9] Does the growing number of scientists, patents, and doctoral degrees mean that mankind is getting continuously wiser? On a *technical* level, scientific development continues and will continue further. We will have electric cars, electric airplanes, easier space travel in the near future, better vaccines and drugs, and more. But this is only the *technical* side. All scientists live in parallel in two worlds: the world of science, and this physical world. They marry, have children, earn money, make a career, and eventually retire. In the world of science, scientists live in their cosmos, popularly known as the ivory tower, separated from other areas of reality, including the wild west of marketing, the military world, regulatory agencies, the clinical medical world, and more. Science and scientific organizations perceive themselves as objective and truth-seeking, in constant fight against hidden attempts to suppress science and to promote ideas and theories that instead serve hidden conflicts of interests. Although this is probably true within many specific cosmos of science, we also need to look at the real-world impact of science in many different dimensions, including communication, warfare, transport, medicine, space travel, economics, and many more, and how this in return feeds back into science itself. We are approaching a nodal point in the history of science where on the level of detailed research progress is advancing, but where there are blurs at the interface between the structures of science and society.

Being a scientist has become a recognized career path, with all the advantages and disadvantages that career paths entail. Science appears to allow one to isolate oneself from worldly influences to a certain extent. But science needs also money. Never in human history has so much money been spent on research as today. Science, like many other areas of society, loves to be lulled into the illusion that it hovers over "worldly" interests.

Scientific rules are adhered to in detail. Scientists can leave the connection to the rest of the world to other structures, responsibilities, theories, and justifications. Self-illusions of scientists and science are the natural target of forces that derive their benefit from this weakness in self-perception. Exactly this happened in the global fight against the COVID-19 pandemic. This pandemic is shaking the foundations of today's world. It is normal that at first everyone thinks these foundations are the only thing that can give inner support in a time of deep uncertainty. But it is also time to address the weaknesses in these foundations and to reflect on how the free world can play on its strengths.

Institutions dedicated to science depend on public goodwill if they are funded by the state. The more secure their funding, the easier they attract talented researchers. They all are neutral in theory, but their existence depends also on how they are publicly perceived. A 100 years ago, depression was called "melancholy" and was seen as an expression of inner conflicts that could be traced back to childhood experiences. There were no effective drugs for depression. Depression in young people was considered non-existent.[29] Women and people of color, and of course women of color, in particular, were considered incapable of completing a university degree. Women who worked as teachers had to stay at home as soon as their pregnancy became apparent. Changes in these issues did not from scientific discussions but from changes in people's lives and people's ideas.

Scientists discuss their findings and theories in journals, books, conferences, and more. Careers in the scientific world happen in a competitive environment. Callings to higher academic ordinations are based on the publications.

Science and specifically the life sciences are in transition toward a new paradigm shift where people are less prepared to trust authorities blindly. Instead, they move toward an increased understanding of the role and the limitations of the professions involved in science and healthcare. Mainstream science is barely aware of this paradigm shift and is to a considerable degree still caught up in an authority that used to be associated with higher education and institutional positions. Science, healthcare and the involved institutions are so complex that we cannot analyze all. Instead, a few examples are in the following.

2 Science is part of society

Slavery was not the invention of a few evil politicians, generals, or entrepreneurs. Instead, slavery has accompanied mankind since the beginning of civilization. The transatlantic slave trade between the 1500s and 1800s was not only particularly brutal.[30] The outstanding characteristic of slavery in these centuries of European expansion was its organization on a worldwide industrial level, using modern ships of increasing size. Most ships going to Africa and the Americas were private vessels engaged in the triangular trade that transported guns and manufactured goods to Africa; from there slaves to the Americas; and finally dyes, drugs, sugar, silver, and gold back to Europe.[31] To gain access to Africa and the Americas, scientists had often to hitch rides on slave ships. In the countries they visited, they had often to rely on slavers for food, shelter, mail, equipment, and local transport. When Newton developed his theory of gravity, he studied among many other things also ocean tides and the underlying gravitational tug of the moon. One crucial set of readings came from French slave ports in Martinique. Newton could not have used these data without the slave trade.[32]

Since the 19th century, medicine was regulated more by normality than with health; biomedical concepts became structured around the polarity of normal as opposed to pathological. At the end of the 18th century, the rise of modern nationalism in Europe was linked to middle-class norms of the body and sexuality, establishing a link between the moral and physical (and later psychological) health of the individual and the health of the state and the equation of good health with social conformity. Medical knowledge was among the tools of colonial conquest. Physicians, surgeons, and pharmacists saw diagnosis and treatment as a contest over civilization alongside health and disease.[21] Not all scientists were slave traders in the 18th century. They were not. But science and the slave trade were part of the same increasingly international human society that continuously expanded its worldview. Over time, slavery, the definition of homosexuality as a disease, and numerous other positions that in today's perspective were utterly flawed have been corrected, but not primarily through scientific publications.

Charles Darwin published his ideas on evolution by natural selection in "On the Origin of Species" in 1859. This still caused shock waves when the journal "Nature" was established in 1969.[33] In 1871, Darwing delivered with "The Descent of Man" another bombshell, opening a century and a half of the debate about what it means to be human.[34]

From around the middle of the 20th century, several scientists began trying to understand the social function of science. The book "The social function of science" by JD Bernal was for a while quite influential,[35] although Bernal's influence began to decline with his unlimited defense of the Soviet Union with its totalitarian rule governed by the communist party. Many others also began to explore the nexus between society and science. The term "science of science" emerged, and the debates about "science of science" continue.[36–39] Many of the participating scientists had very moral and moralistic ideas about science. Today's publications on "science of science" reflect more the self-picture of scientists, describing it as a quantitative understanding of the interactions among scientific agents across diverse geographic and temporal scales, in the hope to provide insights into the genesis of scientific discovery, with the ultimate goal of developing tools and policies that have the potential to accelerate science, based on big data-based capabilities for empirical analysis.[40] Science is described as a form of accumulated knowledge through the effort of humans to understand our universe and sees the main difference to the past that today science is advanced by many researchers from different disciplines.[41] But to understand the role of science in society, it is not sufficient to analyze statistically modern publications.

3 The special edition on 150 years "Nature" in 2019

To celebrate 150 years of "Nature," the journal published a collection of reviews and perspectives to allow a snapshot of areas in which in the editors' view science might help ease the transition to a "healthy, sustainable and inclusive future." In the introduction, the authors (all are "Nature" senior editors) refer to the 17 United Nations (UN) sustainable development goals (SDGs).[42-48]

How can the journal "Nature," often described as the world's leading science journal,[33] refer to the 17 UN sustainable development goals (SDGs)? These 17 UN SDGs succeeded the earlier declared eight United Nations (UN) Millennium Development Goals of 2000 (see Table 1).[49,50] The 17 UN SDGs are listed below in Table 2. They were set up in 2015 by the UN General Assembly to be achieved by 2030 as part of the UN Agenda 2030.[43]

Compared to the Millennium Development Goals of 2000, the Agenda 2030 with its 17 SDGs has become even more flowery, extensive, and meaningless. The United Nations (UN), the WHO, and all other subinstitutions of the UN do not represent a world government. They are a platform on which the governments of the countries of the world negotiate, declare, and publish official pronouncements and development goals that, however, are non-binding. They act as a large fig leaf that documents that the world is doing "something." It is *psychologically* understandable that the editors of "Nature" refer to an international collection of goals. But this is not science.

TABLE 1 UN millennium development goals.

1. To eradicate extreme poverty and hunger
2. To achieve universal primary education
3. To promote gender equality and empower women
4. To reduce child mortality
5. To improve maternal health
6. To combat HIV/AIDS, malaria, and other diseases
7. To ensure environmental sustainability
8. To develop a global partnership for the development

TABLE 2 UN sustainable development goals (Agenda 2030).

1. End poverty in all its forms everywhere
2. End hunger, achieve food security and improve nutrition, and promote sustainable agriculture
3. Ensure healthy lives and promote well-being for all at all ages
4. Ensure inclusive and equitable quality education and promote lifelong learning opportunities for all
5. Achieve gender equality and empower all women and girls
6. Ensure availability and sustainable management of water and sanitation for all
7. Ensure access to affordable, reliable, sustainable, and modern energy for all
8. Promote sustained, inclusive, and sustainable economic growth, full and productive employment, and decent work for all
9. Build resilient infrastructure, promote inclusive and sustainable industrialization, and foster innovation
10. Reduce inequality within and among countries
11. Make cities and human settlements inclusive, safe, resilient, and sustainable
12. Ensure sustainable consumption and production patterns
13. Take urgent action to combat climate change and its impacts
14. Conserve and sustainably use the oceans, seas, and marine resources for sustainable development
15. Protect, restore, and promote sustainable use of terrestrial ecosystems, sustainably manage forests, combat desertification, and halt and reverse land degradation and halt biodiversity loss
16. Promote peaceful and inclusive societies for sustainable development, provide access to justice for all and build effective, accountable, and inclusive institutions at all levels
17. Strengthen the means of implementation and revitalize the Global Partnership for Sustainable Development

The UN was established after World War II as the successor of the League of Nations.[51] World War II had changed fundamentally the social structure of mankind. The UN was established to foster international cooperation and prevent future conflicts. The (victorious) powers China, France, the Soviet Union, the UK, and the US became the permanent members of its Security Council. The Soviet Union and the United States emerged as rival superpowers, setting the stage for the Cold War. After the Cold War, the Soviet Union imploded. Its official successor is Russia, but the economic weight of Russia is very limited. Nevertheless, it still has the nuclear armamentarium built up during the Cold War. Russia is sometimes described as an Upper Volta with nuclear weapons (the former Upper Volta has meanwhile become Burkina Faso). In recent years, China has established itself as the new challenger to the military and economic superpower that continues to lead the USA.[52,53]

The UN does not represent a world government, despite the wishes of many scientists and despite the world of diplomacy. The UN undoubtedly has a function in this world, but we must not be lulled into illusions that take the diplomatic language of the UN and other international institutions and structures at face value.

One can hardly contradict the 17 UN sustainable development goals (SDG) in their generality. Who would dare to speak out against the eradication of poverty on earth (goal 1), or the promotion of the wellbeing of all ages? Unfortunately, these 17 goals are nothing but cheap propaganda in a world where governments just do not agree on the goals of their actions. Russia occupies peripheral areas of neighboring countries and areas of other countries. Its crimes against humanity that are committed in many countries are neither eliminated nor made better by the 17 UN SDGs. As mentioned in a previous chapter, the WHO definition of health in the Declaration of Alma Ata of 1978 defines health as "a state of complete physical, mental and social wellbeing, and not merely the absence of disease or infirmity."[54] Only a few billionaires would be healthy by this definition, which, unfortunately, has little to do with reality. But it reflects the level that government officials can agree on.

4 The "Nature" coverage of the European Union (EU) science budget deliberations in 2019

For the next decade, action on climate change is soaring up the EU's legislative agenda. It is aiming to cut its greenhouse gas emissions by at least 40% (from 1990 levels) by 2030. Last year, the commission proposed that €320 billion, one-quarter of its proposed 2021–27 spending package, be spent on meeting those targets—up from €206 billion (one-fifth) of the current budget. More than one-third of the Horizon Europe budget is supposed to be committed to this effort. A paper from "Nature" in 2019 describes how the EU parliament and the leaders of its member states made record time when they reached an agreement that would supply researchers in Europe with more than €100 billion over 7 years.[55] The key characteristic of this article is that it is not the report of a newspaper, but is written by two journalists who were allowed to publish their story in "Nature." To a considerable degree, the journal "Nature" has degenerated into a lobbying institution for as much money as possible for research.

An article with the catchy title "Europe the rule-maker" shows a European Space Agency (ESA) Ariane rocket ferrying four EU Galileo navigation satellites into space.[45] The picture itself may look nice, but Ariane's technology has in the meantime been surpassed by various US companies who allow space flight at a fraction of the costs.[56–60] The editors of "Nature" accepted this picture, which documents the Stone Age technology with which the EU dreams of being a world leader. The challenge of Europe's Ariane rocket and space flight industry, in general, is usual for economic processes in which, for political reasons, companies and technologies are supported that scientifically and technically are already outdated. Today's rockets of the US Company SpaceX return to the launch site autonomously, can be re-used, and thus are pushing down the price for space flight. In contrast, the Ariane 6, which is now expected to fly a first time in 2022, is still characterized by the principle that nothing will be reused. This worked well half a century ago. But times are changing.

A paper on "Europe the rule-maker" outlines that the EU wants to lead the world's approach to a host of policy agendas informed by science, including climate change, chemicals regulation, and data protection. It further outlines that Europe is proactively filling a void in international scientific leadership created by the US retreat from multilateralism under President Donald Trump, affecting science and many more spheres. It claims that through its emphasis on research that is "cosmopolitan, open and mission directed," the EU is "undeniably" in the driving seat of global scientific governance. It emphasizes the author's hope that the next European Parliament and European Commission should ensure that research and its good governance remain at the top of their agendas. Allegedly, this is one arena in which Europe now leads, and others follow.[45]

Also in 2019, "Nature" asked several leading Europeans to pick their top priority for science at this pivotal point. Carlos Moedas, European Commissioner, Research, Science and Innovation, proudly declares "I strongly believe that Europe must lead the fightback—to encourage our societies to do as climate activist Greta Thunberg urges, and "Listen to Science!" Furthermore, he proudly announces the EU policy of "open-science," i.e., open source access of all publications that receive public funding.[61] But the UN and social media theater around Greta Thunberg have a limited relationship with true science. And to have all publications supported by EU funding be immediately accessible open source would only make a difference if the quality of these papers would make them a contribution to advance science. Unfortunately, given the propaganda in "Nature," the hope for high scientific content of these publications is rather limited.

There are more papers and voices in the Nature 2019 collection that outline how Europe is allegedly a globally leading force in science, development, and more.[45,55,61–64] Nice visions and dreams, but not strongly connected to reality.

Yet another article describes that EU research programs encourage scientists to collaborate across borders and offer lessons for the rest of the world and claims that the value of an integrated EU for science has not been lauded enough. The article describes which lessons the new EU parliament, and others interested in furthering global research should take from EU history. Rigorous research requires collaboration, long-term planning, and stability, and holding fast to those strengths against populism, nationalism, or any other forces that threaten to drive countries apart. The future parliament should champion this, and individual researchers can, too. They should simply tell others how the EU is fundamental to their work.[63] As outlined above, being a scientist has several sides. One is simply to get a salary and be accepted and praised by other scientists. But that alone is no guarantee for good science. Like all professions, also scientists have to face the temptation of trade-union-like demands for good salaries and job security. But such demands do not translate automatically into good science.

5 The term "medical-industrial complex" and its ramifications

In the mid-1960s a group of New York activists founded the Health Policy Advisory Center (Health/PAC) and edited a monthly bulletin with a "New Left" perspective on health. In 1969, Health/PAC first used the term "medical-industrial complex," characterizing the US health system with a term that was a spin-off from President Eisenhower who in his farewell address in 1961 had discussed the dangers of the "military-industrial complex," a term that tries to characterize the relationship between a nation's armed forces and the defense industry, seen together as a force that influences public policy. In the ensuing discussion, the term "military-industrial complex" was described as a force that attempts to marshal political support for continued or even increased military spending by the national government.[65] The term gained popularity in the US after 1961.[66–70]

The term "medical-industrial complex" was first time used in 1969 in the Health/PAC Bulletin,[71] then in a book in 1970.[72] This term and concept were widely discussed throughout the 1970s, including a review of the book in the New England Journal of Medicine (NEJM),[73] and in a paper by Relman, then editor of the NEJM, in 1980. He commented that the past decade had seen the rise of another kind of private "industrial complex" with an equally great potential for influence on public policy in health care. In his article he explicitly excluded pharmaceutical companies from criticism: "They have been around for a long time, and no one has seriously challenged their social usefulness. Furthermore, in a capitalistic society, there are no practical alternatives to the private manufacture of drugs and medical equipment." Instead, he listed proprietary hospitals, home care, laboratory and other services, and hemodialysis. One key question he asks in the discussion is "Can we leave health care to the marketplace?" Even if we believe in the free market as an efficient and equitable mechanism for the distribution of most goods and services, there are many reasons to be worried about the industrialization of health care.[74] John Ehrenreich published a follow-up book in 2014: "Third Wave Capitalism: How Money, Power, and the Pursuit of Self-Interest Have Imperiled the American Dream."[75] In contrast to the differentiated analysis by Relman, criticism of the healthcare business was and is expressed in an almost uncountable number of books that accuse the pharmaceutical industry of almost everything bad under the sky and equalize it to organized crime.[76-87] Obviously, there is a market for readers who enjoy reading healthcare business condemnations, and there are several writers who enjoy serving that market.

Both the criticism of the military-industrial complex and the medical-industrial complex are superficial and short-sighted. The USA as a country survived the Cold War, while the Soviet Union imploded. Today's life science industry allows the treatment of diseases that were formerly either a death sentence, such as relapsing/remitting ALL,[88-90] or that brought lifelong agony of pain and severely impaired quality of life. Without the allegedly "medical-industrial complex," the development of effective vaccines against COVID-19 would not have been possible.[91]

The fundamental criticism of pharmaceutical medicine came in part from respected academic writers, including Marcia Angell, former editor-in-chief of the NEJM.[77,92] But beyond and in contrast to the moralistic and superficial criticism of the pharmaceutical industry, which has dominated the public discussion about modern health research for decades, a different discussion has emerged in academic medicine. Jeffrey Drazen formulated the provocative opinion that there is no real scientific reason not to let experts who work in the industry write editorials or review articles.[93] This editorial was followed by three editorials by Lisa Rosenbaum, who analyzed the potential relationship between high-level medical journals such as the NEJM and people who work in the pharmaceutical industry. Essentially, the conclusion of all these three editorials is that the key issue is not if somebody receives a salary from a pharmaceutical company, but the extent to which their writing make medical and scientific sense.[94-96]

These editorials caused three former senior NEJM editors to respond with dismay in the *British Medical Journal* (BMJ). Essentially, they claim that there are two types of authors for medical journals: those without conflicts of interest (which they include themselves) and those with conflicts of interest, including all connected somehow with the pharmaceutical industry. Only the white knights who have bathed in the holy water of freedom from conflicts of interest should be allowed to write in respected medical journals. This construct is ridiculous. It uses the trust of the public toward the clinical world but represents simply another conflict of interest: a professional group that tries to defend its reputations against the real world.[97]

Interestingly, one of the three former senior NEJM editors published earlier a paper that directly uses the high trust of the public in the clinical profession that is allegedly free of conflicts of interest.[98] The key challenge of the entire "pediatric drug development" is the blur between the administrative and the physiological meaning of the term "child." We associate with "children" spontaneously small children. Most people will welcome the additional testing of medicines in immature young children where in the past toxicities have been observed when drugs were administered. But the result of the "pediatric drug development" movement was that separate "pediatric" studies in all administratively defined young people are required by the authorities, and are paid by pharmaceutical companies. Ironically, among those who benefit(ed) from these studies are also academic clinicians that are supposedly free of conflicts of interest. The pharmaceutical industry paid and continues to pay for these studies that were and are required by the regulatory authorities. But young people grow up physically long before their 18th birthday. Nevertheless, young, physically mature people are included in "pediatric" studies. Such studies are completely unnecessary for adolescents. But they are an elegant way to use pharmaceutical industry money to fund pseudo-scientific academic research.[99,100]

6 Conclusions

Life science industries, publishing, clinical care, and academic medicine are facets of today's world of health care in which an increasing proportion of people do not work directly in the clinical care of patients, but in areas devoted to research, coordination, and communication. In addition to the professions of priests and warriors, direct patient care is one of the oldest and most traditional professions of the mankind. It is understandable that representatives of an old profession dream of representing the true doctrine and of being morally and scientifically superior to all others. But it is also laughable. It is a cruel but simple truth that representatives of the health professions are largely helpless against contagious diseases as long as there is no effective vaccination. Arrogance and a feeling of superiority are qualities that we have to deal with when dealing with representatives of various health professions. The COVID-19 pandemic was and is a high-tech challenge in which the ability to develop modern vaccines is the only mid-term and long-term way to get back to a reasonably normal world.

The underlying currents of thought that exist today in the question of health care are currents at the intersection of politics, science, law, communication, and other areas. One must, on the one hand, cast the net wide to capture critical elements, and, on the other hand, go into sufficient depth to identify key elements. Because there are so many different channels of communication today, it is confusing at first when you try to form your own opinion. On the other hand, the COVID-19 pandemic is not the first and will not be the last challenge humanity has to face.

References

1. Britannica. *Science*. https://www.britannica.com/science/science.
2. Wikipedia. *Science*. https://en.wikipedia.org/wiki/Science.
3. American Association for the Advancement of Science (AAAS). Mission and History. www.aaas.org/mission.
4. ScienceDirect. www.sciencedirect.com/.
5. *Nature*. https://www.nature.com/.
6. ScienceNews. Independent Journalism since 1921. www.sciencenews.org/.
7. The journal "Science". www.jstor.org/journal/science.
8. Wikipedia. *Scientific Journal*. https://en.wikipedia.org/wiki/Scientific_journal.
9. Kuhn TS. *The Structure of Scientific Revolutions. Enlarged. International Encyclopedia of Unified Science*. 2nd ed. Chicago, IL: The University of Chicago Press; 1970. https://www.lri.fr/~mbl/Stanford/CS477/papers/Kuhn-SSR-2ndEd.pdf.
10. C. Nickson Precautionary Principle and the Kehoe Principle n.d. https://litfl.com/precautionary-principle-and-the-kehoe-principle/.
11. Nriagu JO. Clair Patterson and Robert Kehoe's paradigm of "show me the data" on environmental lead poisoning. *Environ Res*. 1998;78(2):71–78.
12. Wikipedia. *Robert A Kehoe*. https://en.wikipedia.org/wiki/Robert_A._Kehoe.
13. Needleman H.L. History of Lead Poisoning in the World. n.d. https://www.biologicaldiversity.org/campaigns/get_the_lead_out/pdfs/health/Needleman_1999.pdf.
14. Needleman HL. The removal of lead from gasoline: historical and personal reflections. *Environ Res*. 2000;84(1):20–35.
15. Wikipedia. *Herbert Needleman*. https://en.wikipedia.org/wiki/Herbert_Needleman.
16. Riva MA, Lafranconi A, D'Orso MI, et al. Lead poisoning: historical aspects of a paradigmatic "occupational and environmental disease". *Saf Health Work*. 2012;3(1):11–16. https://www.ncbi.nlm.nih.gov/pmc/articles/PMC3430923/pdf/shaw-3-11.pdf.
17. Tisma M. "Rich man, poor man": a history of lead poisoning. *Hektoen Int J*. 2019. https://hekint.org/2019/09/24/rich-man-poor-man-a-history-of-lead-poisoning/.
18. Miracle VA. Lead poisoning in children and adults. *Dimens Crit Care Nurs*. 2017;36(1):71–73.
19. Wikipedia. *DSM*. https://en.wikipedia.org/wiki/Homosexuality_in_DSM.
20. Drescher J. Out of DSM: depathologizing homosexuality. *Behav Sci (Basel)*. 2015;5(4):565–575. https://www.ncbi.nlm.nih.gov/pmc/articles/PMC4695779/pdf/behavsci-05-00565.pdf.
21. Spurlin WJ. Queer theory and biomedical practice: the biomedicalization of sexuality/the cultural politics of biomedicine. *J Med Humanit*. 2019;40(1):7–20. https://www.ncbi.nlm.nih.gov/pmc/articles/PMC6373286/pdf/10912_2018_Article_9526.pdf.
22. PubMed. https://pubmed.ncbi.nlm.nih.gov/.
23. Bornmann L, Mutz R. A bibliometric analysis based on the number of publications and cited references. *J Assoc Inf Sci Technol*. 2015;66(11):28. https://arxiv.org/ftp/arxiv/papers/1402/1402.4578.pdf.
24. Gastfriend E. 90% of All the Scientists That Ever Lived Are Alive Today. Future of Life Institute n.d. https://futureoflife.org/2015/11/05/90-of-all-the-scientists-that-ever-lived-are-alive-today/.
25. Larsen PO, von Ins M. The rate of growth in scientific publication and the decline in coverage provided by science citation index. *Scientometrics*. 2010;84:575–603. https://www.ncbi.nlm.nih.gov/pmc/articles/PMC2909426/pdf/11192_2010_Article_202.pdf.
26. Wikipedia. *Royal Society*. https://en.wikipedia.org/wiki/Royal_Society.
27. *The Expansion of Modern Science and Technology*. https://archive.unu.edu/unupress/unupbooks/uu09ue/uu09ue07.htm.
28. Wikipedia. *OECD*. https://en.wikipedia.org/wiki/OECD.
29. Varley CK. Treating depression in children and adolescents: what options now? *CNS Drugs*. 2006;20(1):1–13.

30. Duquette NJ. Revealing the relationship between ship crowding and slave mortality. *J Econ Hist*. 2014;74(02):535–552.
31. Wikipedia. *Triangular Trade*. https://en.wikipedia.org/wiki/Triangular_trade.
32. Kean S. Science's debt to the slave trade. *Science*. 2019;364(6435):16–20.
33. *History of Nature*. https://www.nature.com/nature/about/history-of-nature.
34. Ball P. Science must move with the times. *Nature*. 2019;575(7781):29–31.
35. Bernal JD. *The Social Function of Science*. Georg Rouledge & Sons; 1939.
36. Pielke R. In Retrospect. The social function of science. *Nature*. 2014;507:427. https://www.nature.com/articles/507427a.pdf.
37. Mackay AL. J D Bernal (1901-1971) in perspective. *J Biosci*. 2003;28(5):539–546. https://www.ias.ac.in/article/fulltext/jbsc/028/05/0539-0546.
38. Brown A, Mackay AL. J D Bernal and the replication of the genetic material—hindsight on foresight. *J Biosci*. 2005;30(4):407–409. https://www.ias.ac.in/article/fulltext/jbsc/030/04/0407-0409.
39. Brownhill R, Merricks L. Ethics and science: educating the public. *Sci Eng Ethics*. 2002;8(1):43–57.
40. Fortunato S, Bergstrom CT, Börner K, et al. Science of science. *Science*. 2018;359(6379), eaao0185. https://science.sciencemag.org/content/359/6379/eaao0185/tab-pdf.
41. Zeng A, Shen Z, Zho J, et al. The science of science: from the perspective of complex systems. *Phys Rep*. 2017;714–715:1–73. https://reader.elsevier.com/reader/sd/pii/S0370157317303289?token=F1CBDC31700694263DC778C98204533007CFB19A26C5B1E2BE6673B540500BDAF82261608F2958E1412F85BBD82557FE&originRegion=eu-west-1&originCreation=20210727082458.
42. Armstrong A, Daw R, Hansell C, et al. Science and society. *Nature*. 2019;575(7781):63.
43. United Nations. *17 Goals to Transform our World*. https://www.un.org/sustainabledevelopment.
44. Wikipedia. *Sustainable Development Goals*. https://en.wikipedia.org/wiki/Sustainable_Development_Goals.
45. Wilsdon J, de Rijcke S. Europe the rule-maker. *Nature*. 2019;569(7757):479–481.
46. Gasparini M, Tarquini D, Pucci E, et al. Conflicts of interest and scientific societies. *Neurol Sci*. 2020;41(8):2095–2102.
47. Haque W, Minhajuddin A, Gupta A, et al. Conflicts of interest of editors of medical journals. *PLoS One*. 2018;13(5), e0197141.
48. Smith A, Sanders JKM. Launching science, society and policy. *R Soc Open Sci*. 2021;8(3):210438. https://www.ncbi.nlm.nih.gov/pmc/articles/PMC8074978/pdf/rsos.210438.pdf.
49. *United Nations Millennium Development Goals*. https://www.un.org/millenniumgoals/.
50. Wikipedia. *Millennium Development Goals*. https://en.wikipedia.org/wiki/Millennium_Development_Goals.
51. Wikipedia. *League of Nations*. https://en.wikipedia.org/wiki/League_of_Nations.
52. Simon R. 'Upper Volta with gas'? Russia as a semi-peripheral state. In: *Globalization and the 'New' Semi-Peripheries*. Palgrave Macmillan; 2009:120–137.
53. Lewis S, Hunnicutt T. *Analysis—Biden talks down Russia, spurs allies in bid to back Putin into a corner*. Reuters; June 17, 2021. https://www.reuters.com/article/us-usa-russia-summit-biden-analysis-idCAKCN2DT0B4.
54. WHO. *Declaration of Alma-Ata*; 1978. https://www.who.int/publications/almaata_declaration_en.pdf.
55. Abbott A, Schiermeier Q. How European scientists will spend €100 billion. *Nature*. 2019;569(7757):472–475.
56. Smith R. *Europe's New Space Rocket Is Incredibly Expensive. And the Higher the Cost Goes, the Less Profitable This Rocket Becomes*. The Motley Fool; November 10, 2020. https://www.fool.com/investing/2020/11/10/europe-space-rocket-incredibly-expensive-airbus/.
57. Parsonson A. *ESA Requests €230 Million More for Ariane 6 as Maiden Flight Slips to 2022*. SpaceNews; October 29, 2020. https://spacenews.com/esa-request-230-million-euros-more-for-ariane-6-as-maiden-flights-slips-to-2022/.

58. Henry C. *Ariane 6 is Nearing Completion, But Europe's Work is Far from Over*. SpaceNews; August 15, 2018. https://spacenews.com/ariane-6-is-nearing-completion-but-europes-work-is-far-from-over/.
59. Fouquet H. *Europe's Ariane 6 Rocket Is Doomed Even Before First Flight, Auditor Finds*. Bloomberg News; February 6, 2019. https://www.bnnbloomberg.ca/europe-s-ariane-6-rocket-is-doomed-even-before-first-flight-auditor-finds-1.1209863.
60. Ralph E. *SpaceX's Reusable Falcon Rockets Have Europe Thinking Two Steps Ahead*. Teslarati; June 29, 2020. https://www.teslarati.com/spacex-reusable-falcon-rockets-europe-response/.
61. Moedas C, Vernos I, Kuster S, et al. Views from a continent in flux. *Nature*. 2019;569(7757):481–484.
62. Schiermeier Q. Europe is a top destination for many researchers. *Nature*. 2019;569(7757):589–591.
63. Editorials. What Europe is doing right. *Nature*. 2019;569:455.
64. Gibney E. Europe's controversial plans to expand defence research. *Nature*. 2019;569(7757):476–477.
65. Wikipedia. *Military-Industrial Complex*. https://en.wikipedia.org/wiki/Military%E2%80%93industrial_complex.
66. Britannica. *Military-Industrial Complex*. https://www.britannica.com/topic/military-industrial-complex.
67. B3 | *Healthcare in the USA: Understanding the Medical-Industrial Complex*. https://phmovement.org/wp-content/uploads/2018/07/B3.pdf.
68. Fallows J. The military-industrial complex. *Foreign Policy*. 2002;133:46.
69. Churchill LR, Perry JE. The "medical-industrial complex". *J Law Med Ethics*. 2014;42(4):408–411.
70. Jupiter J, Burke D. Scott's parabola and the rise of the medical-industrial complex. *Hand (N Y)*. 2013;8(3):249–252. https://www.ncbi.nlm.nih.gov/pmc/articles/PMC3745238/pdf/11552_2013_Article_9526.pdf.
71. Ehrenreich B, Ehrenreich J. *The Medical Industrial Complex*. Bulletin Health Policy Advisory Center; 1969.
72. Ehrenreich B, Ehrenreich J. *The American Health Empire. Power, Profits and Politics. A Report from the Health Policy Advisory Center (Health-PAC)*. NY, USA: Random House; 1970.
73. Ehrenreich B, Ehrenreich J. *Book Review: The American Health Empire. Power, Profits and Politics. A Report From the Health Policy Advisory Center (Health-PAC)*. New York: Vintage Books; 1971:279. 1.95.
74. Relman AS. The new medical-industrial complex. *N Engl J Med*. 1980;303(17):963–970. https://www.nejm.org/doi/pdf/10.1056/NEJM198010233031703?articleTools=true.
75. Ehrenreich J. *Third Wave Capitalism: How Money, Power, and the Pursuit of Self-Interest Have Imperiled the American Dream*. Ithaca, NY: Cornell University Press; 2016.
76. Wikipedia. *Medical–Industrial Complex*. https://en.wikipedia.org/wiki/Medical%E2%80%93industrial_complex.
77. Angell M. *The Truth About the Drug Companies: How They Deceive Us and What to Do About It*. New York, USA: Random House Publishers; 2004.
78. Whitacker R. *Mad in America: Bad Science, Bad Medicine, and the Enduring Mistreatment of the Mentally Ill*. revised ed. New York, USA: Basic Books; 2019.
79. Soloway S. *Bad Medicine: The Horrors of American Healthcare*. New York, USA: Skyhorse; 2020.
80. Grouse L. Cost-effective medicine vs. the medical-industrial complex. *J Thorac Dis*. 2014;6(9):E203–E206. https://www.ncbi.nlm.nih.gov/pmc/articles/PMC4178073/pdf/jtd-06-09-E203.pdf.
81. Cassels A. Ponziceuticals within the medical industrial complex. *CMAJ*. 2009;181(1–2):112. https://www.ncbi.nlm.nih.gov/pmc/articles/PMC2704400/pdf/1810112.pdf.

82. Posner G. *Pharma: Greed, Lies, and the Poisoning of America*. New York, USA: Avid Reader Press/Simon & Schuster; 2020.
83. Laskai A. *Institutional Corruption Theory in Pharmaceutical Industry-Medicine Relationships: A Qualitative Analysis of Hungary and the Netherlands (Studies of Organized Crime)*. Cham, Switzerland: Springer Nature; 2021.
84. Rosenthal E. *An American Sickness: How Healthcare Became Big Business and How You Can Take It Back*. Penguin Press; 2017. ISBN:9780698407183.
85. Gøtzsche PC. *Deadly Medicines and Organised Crime: How Big Pharma Has Corrupted Healthcare*. London, UK: Routledge; 2013.
86. Goldacre B. *Bad Pharma: How Drug Companies Mislead Doctors and Harm Patients*. London, UK: Harper Collins Publishers; 2012.
87. Goldacre B. *Bad Science: Quacks, Hacks, and Big Pharma Flacks*. NY, USA: Farrar, Straus and Giroux (FSG) Adult; 2010. reprint.
88. Maude SL, Laetsch TW, Buechner J, et al. Tisagenlecleucel in children and young adults with B-cell lymphoblastic leukemia. *N Engl J Med*. 2018;378:439–448.
89. *Emily Whitehead: A Young Girl Beats Cancer with Immunotherapy*. https://www.cancerresearch.org/immunotherapy/stories/patients/emily-whitehead.
90. Emily Whitehead Foundation. *2020 Stem Cell & Regenerative Medicine Action Award honoree*. https://www.youtube.com/watch?v=GW42FUhflkY.
91. Slaoui M, Hepburn M. Developing safe and effective Covid vaccines—operation warp Speed's strategy and approach. *N Engl J Med*. 2020;383:1701–1703.
92. Angell M. Industry-sponsored clinical research: a broken system. *JAMA*. 2008;300(9):1069–1071.
93. Drazen JM. Revisiting the commercial–academic interface. *N Engl J Med*. 2015;372(19):-1853–1854. https://www.nejm.org/doi/pdf/10.1056/NEJMe1503623?articleTools=true.
94. Rosenbaum L. Reconnecting the dots—reinterpreting industry-physician relations. *N Engl J Med*. 2015;372:1860–1864. https://www.nejm.org/doi/pdf/10.1056/NEJMms1502493?articleTools=true.
95. Rosenbaum L. Understanding bias—the case for careful study. *N Engl J Med*. 2015;372:1959–1963. https://www.nejm.org/doi/pdf/10.1056/NEJMms1502497?articleTools=true.
96. Rosenbaum L. Beyond moral outrage—weighing the trade-offs of COI regulation. *N Engl J Med*. 2015;372:2064–2068. https://www.nejm.org/doi/pdf/10.1056/NEJMms1502498?articleTools=true.
97. Steinbrook R, Kassirer JP, Angell M. Justifying conflicts of interest in medical journals: a very bad idea. *BMJ*. 2015;350, h2942.
98. Steinbrook R. Testing medications in children. *N Engl J Med*. 2002;347(18):1462–1470.
99. Rose K. *Considering the Patient in Pediatric Drug Development. How Good Intentions Turned into Harm*. London: Elsevier; 2020. https://www.elsevier.com/books/considering-the-patient-in-pediatric-drug-development/rose/978-0-12-823888-2. https://www.sciencedirect.com/book/9780128238882/considering-the-patient-in-pediatric-drug-development.
100. Rose K. *Blind Trust—How Parents With a Sick Child Can Escape the Labyrinth of Lies, Hypocrisy and False Promises*. London, UK: Hammersmith Books; 2021. https://www.chapters.indigo.ca/en-ca/books/blind-trust/9781781612026-item.html. and https://www.amazon.com/-/de/dp/1781612021/ref=sr_1_2?dchild=1&qid=1620137631&refinements=p_27%3AKlaus+Rose&s=books&sr=1-2.

CHAPTER 12

Vaccination hesitancy

1 Introduction

In the past years and decades, there has been a shift in the public and media perception of vaccines, which has facilitated outbreaks such as mumps. Organized opposition to vaccinations has a long history as old as vaccination itself. While in the 19th-century vaccination had initially become compulsory in Britain, antivaccination leagues challenged the law as a violation of civil liberty. Eventually, the vaccination law was amended in 1898 to allow the exemption for parents, based on conscience, which introduced the concept of "conscientious objector" into English law.[1] In the USA, all states and Washington DC (District of Columbia) allow medical exemptions when a child has a medical condition that prevents it from receiving a vaccine. All but three states offer nonmedical exemptions for religious or philosophical reasons.[2]

There are very different types of concerns, and it is better not to lump them all together.[3,4] The decision to have yourself or your children vaccinated is in any case complex and contains many different dimensions. Until recently, the vaccination decision was usually made on children for whom parents have to decide. In the COVID-19 pandemic, it was the other way around for a while, since at the beginning of the pandemic young people very rarely got COVID-19. This is gradually changing as a growing proportion of the adult population is vaccinated; the virus successfully mutates further, and increasingly infects also younger persons. Basically, this does not change the reasons for deciding for or against vaccination.[5,6]

In times of impactful and threatening societal events or developments—such as climate change, economic or financial crises, terrorism, war, or public health problems—many people make assumptions about the deceptiveness and evil intentions of powerful leaders or even entire branches due to their uncertainty and fear. These assumed culprits include the pharmaceutical industry, financial institutes, religions, and more—just the usual suspects ☺.[7] The belief in conspiracy theories has been prevalent throughout human history.[8] Nonetheless, it would be short-sighted to try to explain vaccination hesitancy in the time of COVID-19 only to conspiracy theories.

While potential risks of vaccines are visible to persons who believe vaccines are the cause of their or their child's medical conditions, the benefits of vaccines for individuals and communities are today far more difficult to observe. It has often been stated that vaccines are victims of their success. The diseases that vaccines prevent have become rare and very rare in developed countries; partially due to improved hygiene, nutrition, housing, and many more factors that contribute to today's standard of living; but to a very relevant degree also due to successful vaccination programs. Many parents and young physicians have no first-hand experience of the damage diseases can cause that are today mostly prevented by vaccination, but that only a generation ago caused vast suffering, often lasting damage, and even death.[9]

First, a quick look (again) at our increasingly complex world and the relationship between common sense, our perception, and experiences, and how we come to our own opinion based on reports and stories from conversations, from the social media, and the classic media including newspapers, magazines, TV, radio, books, internet, and more. Matters as simple and banal as having to buckle up in a car or airplane were preceded by bitter arguments that lasted for decades. Nobody regularly experiences one car accident after the other. However, it is different if you work as a traffic policeman, an ambulance driver, or as a doctor, nurse, or cleaning person in the emergency department of a hospital or trauma surgery. Before cars existed, there were no car accidents. As the car spread, so did the number of accidents. For a while, surgical removal of an eye was one of the compulsory operations young medical doctors had to undergo during the training to become an ophthalmologist. As long as there was no obligation to wear seat belts, there was no shortage of patients whose faces had been so severely cut in an accident with broken windshields that an eye removal was inevitable. That changed when the seat belt obligation was introduced and the police enforced it. Today the protest against the obligation to wear seat belts no longer belongs to the essential creeds of the great freedom fighters. Over the past few decades, there has been a growing trust that seat belts, for the most part, serve a useful purpose and prevent avoidable injury. Our l observations have not changed, but the way we process them has.

In health care, the days are over when the individual doctor checks whether a new drug is working. This has been replaced by the complex procedure of drug development including drug approval by the regulatory authorities.[10] The key tests are done in clinical studies which are complicated, expensive, and usually directly involve dozens or even hundreds of test centers with doctors, nurses, and other hospital staff. In addition, the patients first have to reach the trial center, even if they live somewhere in the middle of nowhere. However, in the meantime, home visits by study nurses have become an integral part of conducting clinical trials. It would not be feasible to relocate a whole trial to a patient's home. But it is possible to combine on-site visits with at-home visits. Body assessment, vital signs, blood collection, questionnaire collection, administration of the investigational product, and several more procedures can be performed by a trained nurse at home.[7] In addition, every study is monitored to ensure that not a single patient is fictitious, that the consent of each individual patient to participate is well documented, that the inclusion criteria for each individual patient have been correctly checked, and more. The responsible company itself or the clinical trial organization that has been entrusted with conducting the study performs these monitoring procedures themselves. Furthermore, the authorities do the same on a random basis. If they find anything fishy or openly wrong, the whole study can be invalidated. Then it will have to be repeated, which will cost a lot of money and precious time. Therefore, the developing pharmaceutical companies themselves have a great interest in ensuring that everything is absolutely solid.

To the extent that our world becomes more complex and more and more people are involved in ever more complex processes, and that the respective administration of the associated machinery becomes increasingly bureaucratized, several things happen in different dimensions. First, the individual has no overview of the many individual aspects of every single study, in the development of a drug, or its approval. Secondly, it happens occasionally that individual components of the described machinery are spreading false information, be it for criminal, psychiatric, religious, political, or other reasons. This can unfold a disastrous effect since the general public is faced with a world that is difficult to understand. Thirdly, things keep coming to light that shows that institutions and authorities that have long been almost completely trusted have done criminal things whose wickedness and meanness are inconceivable according to our present-day ideas, including the worldwide involvement of Christian and other authorities in the mistreatment of orphans, indigenous children, and single mothers who were unable to defend themselves.[11–13] Fourth, it still happens today that trustworthy institutions disseminate false information, lie openly, or more or less consciously disseminate incorrect information. It does not have to be just individual politicians who are trusted to be lying anyway. Which government has so far openly admitted that the previous management of the COVID-19 pandemic was rather a series of wrong decisions? All governments claim to have done everything right. Anyone who criticizes this in China or Russia ends up in prison or disappears. In the free world, you can at least open your mouth. There are many good reasons to be suspicious of official information. The concerns people have about vaccinations and/or other government-controlled health measures are also evidence that people carry with them survival instincts and common sense from prehistoric times. But that does not mean that people who doubt the safety or effectiveness of the COVID-19 vaccines are right.

Vaccination hesitancy is not just the result of an ongoing war on science as some authors write.[14,15] There are some, and unfortunately more than a few weirdos who wage war like this. But they do not appeal to idiots. They cleverly put a mixture of true, fake, and other information into contexts where it is not easy to see through misinformation. It is good that people are also suspicious. Building trust takes a long time and a lot of effort. But trust is also quickly destroyed. Vaccination hesitancy is, to some extent, the result of a loss of trust in the health system. The reasons for this are not just false claims or because people are deceived. Other equally weighty reasons include the arrogance of health professionals, hidden conflicts of interest being swept under the carpet, and more. In the following, we go through several reasons that may cause vaccination hesitancy. If you are a doctor and want to convince a patient, you will reach most of them with reasonable argumentation, but not all. If you want to convince a friend or colleague of the usefulness of the COVID-19 vaccination, good arguments will convince many, but not all.

The most well-known example of a false claim against vaccination is the claimed link between autism and the measles, mumps, and rubella (MMR) immunization, which is frequently reported and discussed in the literature.[15–17] Andrew Wakefield is a British anti-vaccine activist, former physician, and discredited academic. When he was still a medical doctor, he attracted attention in 1993 when he claimed a link between the measles virus and Crohn's disease.[18] Crohn's disease is a chronic severe inflammatory bowel disease that can affect any part of the gastrointestinal tract from the mouth to the rectum and the anus. Symptoms can include abdominal pain, diarrhea, bleeding from the bowel, fever, and weight loss. In around a quarter of patients, Crohn's disease begins in childhood or teenage years. In 1995, Wakefield claimed a link between the measles vaccines and Crohn's disease in the prestigious medical journal The Lancet.[19] Both claims could not be confirmed by other researchers. While he was still researching Crohn's disease, he had been approached by the parent of an autistic child who was seeking help for bowel problems. Wakefield turned his attention to possible connections between the mumps, measles, rubella (MMR) vaccine, and autism. In 1998, he published with 12 other colleagues the mentioned paper in The Lancet, in which the authors falsely claimed a link between the measles, mumps, and rubella (MMR) vaccine and autism.[16] The paper itself did not claim a causal connection, but Wakefield had already made such statements at a press conference and in a video news release issued by the hospital, calling for the suspension of the triple MMR vaccine until more research could be done. This was immediately controversial, leading to widespread publicity and a drop in vaccination rates in the UK. It was the beginning of a fear of the MMR vaccination throughout the world. Following Wakefield's claim, many studies were undertaken worldwide to check his allegations. None confirmed them. In 2004, reporter Brian Deer revealed in The Sunday Times that, before submitting his paper to The Lancet, Wakefield had received £55,000 from legal firms seeking evidence to use against vaccine manufacturers; that several of the parents quoted as saying that MMR had damaged their children were also litigants; and that Wakefield had not informed his colleagues, co-authors, or the medical authorities of this conflict of interest. Immediately after these allegations, 10 of his 12 co-authors retracted. In 2007, Wakefield and two of his co-authors were charged by the British General Medical Council, which is responsible for licensing doctors and supervising medical ethics in the UK, of serious professional misconduct. In 2010, the General Medical Council found that Wakefield had been dishonest in his research, had acted against his patients' best interests, had mistreated developmentally delayed children, and had failed in his duties as a responsible medical doctor. The Lancet retracted his publication.[20,21] Wakefield was struck off the UK medical register, which barred him from practicing medicine in the UK. However, fear of autism continues to be frequently cited regarding the MMR vaccine among parents in different settings.[16,22]

There are numerous further examples in the literature for vaccination controversies that have led to decreased vaccination rates and even failure of an immunization program in various developing countries. In Cameroon in 1990, there were rumors and fears that public health officials administered childhood vaccines to sterilize women; this thwarted the country's immunization efforts. In the Philippines in the 1990s, the Catholic Church raised concerns about tetanus immunizations, sparking sterilization rumors and halting the campaign. Another example was the boycott of the polio vaccine in northern Nigeria in 2003, where political and religious leaders of five states brought the immunization campaign to a halt by calling on parents not to allow their children to be immunized due to fears that the vaccine was a Western plot to spread infertility and HIV among Muslims. The boycott was short-lived in most states, but the state of Kano maintained the ban for almost a year, leading to a resurgence of polio in Nigeria that then spread to at least 15 countries that had been previously free of polio. Negative rumors about the polio vaccine had circulated year before, but several historical, cultural, political, and contextual factors contributed to this crisis including past attempts to regulate population and fertility, distrust of the West, and the re-election of a southern President over the northern Muslim candidate.[16,22]

There were and are many studies trying to analyze and understand vaccine hesitancy around the globe. The factors that contribute to the expression and development of vaccination concerns can be almost innumerable.

They can include public health crises such as mad cow disease; the belief that vaccinations are against God's will; the belief that vaccination occurs in the context of general oppression of women; rumors of real and/or fictitious side effects; and many more. The educational background, social status, belonging to a social minority, the gender of the respective parent, and many other factors can play a role.[16,22,23]

Although our world has profoundly changed during the past two centuries; despite many much safer and more effective vaccines having been developed; and despite a much better-organized surveillance of adverse events and side effects after vaccination, vaccine opposition is still as deeply rooted as it was two centuries ago. Some of the arguments of anti-vaccination activists in the 1800s are still used today: vaccines are ineffective or cause diseases; vaccines are used to make a profit; vaccines contain dangerous substances; harms caused by vaccines are hidden by the authorities; vaccination mandates violate civil rights; natural immunity is better than immunity induced by vaccines or natural approaches to health; alternative products such as homeopathy, vitamins or nutritional supplements are superior to vaccines to prevent diseases, and so forth. However, there are also fundamental differences between anti-vaccinists then and now. In the past anti-vaccination activists were mostly poor people who were opposed to the state intervention in their bodies and the bodies of their children. Today's anti-vaccinists in high-income countries are often well-educated middle- and upper-income parents. Many contemporary anti-vaccination groups were formed by parents who believed that their child has been harmed by vaccines and therefore expected and demanded compensation from the industry or the government. Other anti-vaccination groups are led by alternative practitioners who are opposed to research-based biomedicine and who promote "natural solutions" to replace vaccination.[16,17,22,23]

The re-emergence of major measles outbreaks in countries in which it had previously been well controlled or eliminated is the most dramatic example of the deleterious impact of vaccine hesitancy on community protection: a substantial proportion of cases reported in the USA were intentionally unvaccinated.[24]

Some complain that information targeting underserved racial/ethnic minorities in the USA about the need to inform to mitigate SARS-CoV-2 infection and transmission in a culturally competent manner has been lacking so far. They emphasize that this information is crucial for educating these communities about COVID-19 vaccines and their distribution as well as dispelling misinformation regarding vaccine trials, safety, and efficacy. They also claim that the lack of education has greatly contributed to COVID-19 vaccine hesitancy and will increase disparities in vaccine uptake. They emphasize that timely vaccinations are also essential to curtailing virus transmission and the emergence of SARS-CoV-2 variants that may evade the immune response produced by the currently existing COVID-19 vaccines. They discuss that minorities and marginalized communities such as medically disabled, homeless, indigenous peoples, and undocumented immigrants, need better access to the COVID-19 preventive infrastructure. They propose other forms of available assistance to individuals who have no access to vaccination sites, including the use of mobile vaccine delivery units and the deployment of vaccine administrators who deliver vaccines to the homeless and other transient and immobile populations, and reflect on the expansion of vaccination sites to unconventional venues, such as the Department of Motor Vehicles, restaurants, churches, salons, coffee shops, post offices, grocery stores, bus/train stations, theme parks, and shelters.[25] We are back to another fundamental challenge: the balance between what is theoretically desirable, what public health activists are demanding, and what is practically feasible and affordable. There must always be a balance between all of these factors. Finding that balance is never easy, and in the end, it is impossible to please everyone.

There are several aspects regarding vaccine hesitancy in the time of the COVID-19 pandemic. Vaccine hesitancy is nothing new, has been with mankind since the first effective vaccine was introduced, and is unlikely to go away entirely. The concerns raised about vaccination during the COVID-19 pandemic are nothing new. But there are also some new aspects. First, the vaccinations available today have been developed in record time, at a rate never seen before in history. If people are afraid that the vaccines are unsafe for whatever reason, this fear is understandable at first, especially as long as science and the media struggle to explain how the vaccines work to someone who has not studied biochemistry. Second, the internet now gives people access to information that previously was beyond their reach. In the past, you had to read scientific publications in the library. Today it is enough to open the computer. More and more scientific publications are open source, i.e., they can be downloaded and read directly without paying a fee. Third, the COVID-19 pandemic has gripped the entire world economy and caused the greatest economic crisis since Black Friday. Of course, this also shook the general worldview of most people. Above all, the COVID-19 pandemic showed that humanity was not well prepared. Governments paint a picture of themselves that they have done everything right. In reality, the pharmaceutical industry was a major force that made a difference. This would not have been possible without the scientific progress of the last few decades, without governments paying for the vaccines, the regulatory authorities approving the vaccines, the healthcare professionals who administer them, and many more. Nevertheless, in the end, it was not science or scientists who developed and manufactured the vaccines, it was pharmaceutical companies competing with each other. In the previous chapters, we have already discussed several interfaces at which the fight against the COVID-19 pandemic was hooked. The essential steps to combat vaccination hesitancy will have to be in connection with the processing of these interfaces.

2 Discussion

To mentally process vaccination hesitancy in the COVID-19 pandemic, we need to consider different levels. There are different types of vaccination concerns that have almost nothing to do with each other. It is one thing if modern vaccinations in a poor predominantly Muslim country are characterized by some religious leaders as part of a foreign conspiracy, and some people or even the majority believe this. It is an entirely different matter if well-educated persons in a western, affluent country have vaccination concerns. Persons should know from their education and background knowledge that they are putting their children at risk by not vaccinating. And it was a completely different matter when in countries of the European Union, as has happened in Germany, Italy, and other countries, the governments suddenly withdrew the authorization of the effective COVID-19 vaccine developed by AstraZeneca because they followed sensationalistic reports in the media, which reported very rare vaccine side effects. The governments panicked, withdrew their recommendation to vaccinate, and certainly increased the vaccine hesitancy in the general population.

We need to keep in mind that we neither have a world government nor many governments that have kept their nerves in this fundamental crisis of the modern world. There are two major exceptions. One was the decision of the US government to set up Operation Warp Speed. This initiative came from several government institutions and high-level organizations in the public health care system and aimed at developing drugs, vaccines, and diagnostics as soon as possible. Although the same government proved daily that it did not understand the gravity of the COVID-19 pandemic, it was moved to at least one reasonable step by key pillars of public health and public administration. The next president after Donald Trump understood the seriousness of the pandemic. The US had now at least already ordered and paid for enough vaccines so that the vaccination campaign could start fast in 2021. The European Union (EU) has in principle comparable institutions and industry, but its leadership clings to ideas that are decades out of date. Apart from having built an almost uncontrollable administrative monster, they perceive the role of the healthcare industry, including the development of drugs, vaccines, and diagnostics, in the light of outdated ideas and act accordingly. For them, the industry is just greedy and profit-seeking and therefore bad, while medical doctors are white knights and therefore good. But these "white knights" do not understand the development of drugs and vaccines. When it was time to order enough vaccines, the EU simply did not. We will come back to this in the next chapters.

The other country was Israel, where the government ordered enough vaccines in time to vaccinate the entire population early. That cost a lot of money. Israel had to pay high prices. The Prime Minister personally took care of ordering sufficient quantities of vaccines. In contrast to the head of the EU administration, he was not a medical doctor. Instead, he used his common sense. The bottom line is that this investment has certainly paid off economically for Israel. Israel and the US are not homogeneous countries. Orthodox Jews distrust all government regulations. In both countries, members of these groups became infected particularly fast and thoroughly and had a particularly high number of deaths to complain about.

The vaccination hesitancy in affluent countries has a multitude of sides. On the one hand, people in these countries are used to everything working and functioning well, and that infectious diseases almost do no longer occur. In addition to almost everything working anyway, many believe that it is enough to take care of an organic, vegan, or not genetically modified food. Many are spoiled by the affluent society. But all these people vote. Politicians do not want trouble with people who can vote. Therefore, they prefer to talk around the bush.

The other mirror side of the same situation is that people reject reasonable rules of conduct and reasonable recommendations by confusing disobedience and protest with freedom, politics, religion, and other noble thoughts. In our society, it is not forbidden to harm yourself through stupidity.

Governments should not make the situation worse by placing more emphasis on the many rights of citizens than on protecting the majority from the SARS-CoV-2 virus. Increasingly, in countries where the government has taken decisive steps, people can no longer go to restaurants, theaters, or the cinema without a vaccination certificate. Without such a certificate you can no longer take the bus, train, plane, boat, or anything else. After initial protest from the usual freedom fighters, this was accepted by and large. To put it the other way around: people in affluent societies are spoiled, and politicians should not try to make that spoilage worse.

To act in a pandemic like COVID-19, mankind needs a compass. There are more than enough publications on how to treat the disease. Medical doctors have learned a lot. Effective vaccines have been developed. These vaccines reach the population far too late compared to what would technically be possible.

3 Additional dimensions

As the COVID-19 pandemic continues and the availability of effective vaccines increases, we are seeing some more dimensions that were either not visible at the beginning of the pandemic or that were simply not imaginable. Before effective vaccines were available, the idea of effective vaccinations was only a distant dream and one of several possibilities. It could also have been a few years before effective vaccines were available. Now that only a few skeptics are doubting the effectiveness of the vaccinations, some new trends have opened up. The ingredients for the mental constructions with which vaccination skeptics and vaccine opponents operate have expanded.

There is a relevant section among young people who are not very keen to get vaccinated. They have a lot of half-knowledge, hearsay, nonsense from the internet and social media, and other solid sources at their disposal, all of which speak out against vaccinations. Intellectually one cannot take this information and opinions seriously. But it gets serious at the point when young people go to corona parties to get infected because they know very well that young people extremely rarely get COVID-19. In the meantime, the first of these young people, who became infected at such a party and have died, specifically now that new more infectious variants of the virus are around. What is not entirely clear to these young idiots is that they are endangering others through their behavior. And our society has become so used to the basic values of freedom for everyone that it is difficult for politics and the state to stop them. The laws are always based on what happened in the past. Time passes before laws are adapted to new circumstances. During this time, adolescent behavior in young people can do a lot of damage.

A second relevant aspect was and is the mixing of thoughts of freedom with the rejection of contact bans, curfews, wearing masks, and vaccinations. This has led to downright hysterical movements in some countries. It is even worse when such thoughts are then wrapped in a political cloak. And it is even worse when these thoughts are then used by people who have an interest, fun or even a financial interest in doing as much damage as possible.

In our modern, complex society there will always be some people who are skeptical, believe in conspirations or are against almost everything that is done or recommended by the government. That is one of the challenges for politicians: you cannot satisfy everybody. We can only hope that there are enough politicians who use common sense as a basis in their actions and have their own opinion instead of just hanging their cloaks after opinion polls.

4 Conclusions

In order to overcome the COVID-19 pandemic, global vaccination with effective vaccines has already proven to be the decisive step forward. We are not yet there. Even in developed countries, the level of vaccination is not yet satisfactory. One aspect of dealing with spoiled, well-off people who mix up rational and irrational feelings is that it will not be enough to understand them. Good arguments are necessary to convince of the point of vaccinations.[23] But it will also be necessary to demonstrate determination and not overdo it with the understanding. As soon as it becomes more and more difficult to visit restaurants, cinemas, public events and more without a vaccination certificate and as soon as such a certificate is also necessary for the use of trains, buses, airplanes and ships, the shouting of those who oppose vaccination will decrease.

References

1. Wolfe RM, Sharp LK. Anti-vaccinationists past and present. *BMJ*. 2002;325(7361):430–432. https://www.ncbi.nlm.nih.gov/pmc/articles/PMC1123944/pdf/430.pdf.
2. US Centers for Disease Control and Prevention (CDC). *What is an exemption and what does it mean? SchoolVaxView*; 2017. https://www.cdc.gov/vaccines/imz-managers/coverage/schoolvaxview/requirements/exemption.html.
3. Di Pietro ML, Posia A, Teleman AA, et al. Vaccine hesitancy: parental, professional and public responsibility. *Ann Ist Super Sanita*. 2017;53(2):157–162. https://www.iss.it/documents/20126/45616/ANN_17_02_13.pdf.
4. Rosselli R, Martini M, Bragazzi NL. The old and the new: vaccine hesitancy in the era of the web 2.0. Challenges and opportunities. *J Prev Med Hyg*. 2016;57(1):E47–E50. https://www.ncbi.nlm.nih.gov/pmc/articles/PMC4910443/pdf/2421-4248-57-E47.pdf.
5. Etzioni-Friedman T, Etzioni A. Adherence to immunization: rebuttal of vaccine hesitancy. *Acta Haematol*. 2020;1–5. https://www.ncbi.nlm.nih.gov/pmc/articles/PMC7705945/pdf/aha-0001.pdf.
6. Troiano G, Nardi A. Vaccine hesitancy in the era of COVID-19. *Public Health*. 2021;194:245–251. https://www.ncbi.nlm.nih.gov/pmc/articles/PMC7931735/pdf/main.pdf.
7. 'Casablanca' At 75: Five Memorable Quotes That Will Make You Want To Revisit The Classic. The Economic Times Panache; 2017. https://economictimes.indiatimes.com/magazines/panache/casablanca-at-75-five-memorable-quotes-that-will-make-you-want-to-revisit-the-classic/heres-looking-at-you-kid-/slideshow/61793043.cms.
8. Löffler P. Review: vaccine myth-buster - cleaning up with prejudices and dangerous misinformation. *Front Immunol*. 2021;12:663280. https://www.ncbi.nlm.nih.gov/pmc/articles/PMC8222972/pdf/fimmu-12-663280.pdf.
9. Schwartz JL, Caplan AL. Vaccination refusal: ethics, individual rights, and the common good. *Prim Care*. 2011;38(4):717–728. https://www.academia.edu/16650654/Vaccination_Refusal_Ethics_Individual_Rights_and_the_Common_Good.
10. Rägo L., Santo B. Drug regulation: history, present and future. In: van Boxtel C.J., Santo B., Edwards I.R., eds. Drug Benefits and Risks: International Textbook of Clinical Pharmacology, revised 2nd ed. Wiley https://www.who.int/medicines/technical_briefing/tbs/Drug_Regulation_History_Present_Future.pdf.
11. Daly K. Conceptualising responses to institutional abuse of children. *Curr Issues Crim Just*. 2014;26(1):6–29. http://www.austlii.edu.au/au/journals/CICrimJust/2014/10.pdf.
12. Sköld J. The truth about abuse? A comparative approach to inquiry narratives on historical institutional child abuse. *Hist Educ*. 2016;45(4):492–509. https://doi.org/10.1080/0046760X.2016.1177607.
13. Mäkelä D. *Historical Child Abuse in out-of-Home Care: Finland Disclosing and Discussing its Past (Master's Thesis in Child Studies)*. Linköping University; May 2015. http://liu.diva-portal.org/smash/get/diva2:844350/FULLTEXT01.pdf.

14. Adhikari B, Cheah PY. Vaccine hesitancy in the COVID-19 era. *Lancet Infect Dis.* 2021;21(8):1086. https://www.ncbi.nlm.nih.gov/pmc/articles/PMC8248943/pdf/main.pdf.
15. Goldenberg MJ. *Vaccine Hesitancy: Public Trust, Expertise, and the War on Science.* Pittsburgh, PA: University of Pittsburgh Press; 2021.
16. Dubé E, Vivion M, MacDonald NE. Vaccine hesitancy, vaccine refusal and the anti-vaccine movement: influence, impact and implications. *Expert Rev Vaccines.* 2015;14(1):99–117. https://doi.org/10.1586/14760584.2015.964212.
17. Wikipedia. *Vaccination Hesitancy.* https://en.wikipedia.org/wiki/Vaccine_hesitancy.
18. Wakefield AJ, Pittilo RM, Sim R, et al. Evidence of persistent measles virus infection in Crohn's disease. *J Med Virol.* 1993;39(4):345–353.
19. Thompson NP, Montgomery SM, Pounder RE, Wakefield AJ. Is measles vaccination a risk faction for inflammatory bowel disease? *Lancet.* 1995;345(8957):1071–1074.
20. Wakefield AJ, Murch SH, Anthony A, et al. Ileal-lymphoid-nodular hyperplasia, non-specific colitis, and pervasive developmental disorder in children. *Lancet.* 1998;351(9103):637–641.
21. Eggertson L. Lancet retracts 12-year-old article linking autism to MMR vaccines. *Can Med Assoc J.* 2010;182(4):2. https://www.ncbi.nlm.nih.gov/pmc/articles/PMC2831678/pdf/182e199.pdf.
22. Wikipedia. *Andrew Wakefield.* https://en.wikipedia.org/wiki/Andrew_Wakefield.
23. Connell AR, Connell J, Leahy TR, et al. Mumps outbreaks in vaccinated populations-is it time to re-assess the clinical efficacy of vaccines? *Front Immunol.* 2020;(11):2089. https://www.ncbi.nlm.nih.gov/pmc/articles/PMC7531022/pdf/fimmu-11-02089.pdf.
24. Pittet LF, Abbas M, Siegrist CA, et al. Missed vaccinations and critical care admission: all you may wish to know or rediscover—a narrative review. *Intensive Care Med.* 2020;46(2):202–214. https://www.ncbi.nlm.nih.gov/pmc/articles/PMC7223872/pdf/134_2019_Article_5862.pdf.
25. Hildreth JEK, Alcendor DJ. Targeting COVID-19 vaccine hesitancy in minority populations in the US: implications for herd immunity. *Vaccines (Basel).* 2021;9(5):489. https://www.ncbi.nlm.nih.gov/pmc/articles/PMC8151325/pdf/vaccines-09-00489.pdf.

CHAPTER 13

Social inequality, developing countries, and COVID-19

1 Introduction

Social inequality exists in any human society that has left the hunter-gatherer stage behind. The more complex a community was and is, the more there is a division of labor with different tasks in which people can no longer simply take turns, but for which they need different training and professional experience. Differentiated and complex human societies have a hierarchical structure in which the rules of living together and working together are more or less formalized. This goes hand in hand with the fact that those who are higher up in the hierarchy are also generally better off. They live in better houses or accommodations, have access to more and better goods, and more money, since the days that money has existed at all. Today's society is becoming more and more complex.

Social inequality also reflects two fundamentally different characteristics of the human being, some of which contradict, and some of which complement one another. We spontaneously help people who are affected by an accident, lose their house in a tempest, or are affected by any other acts of God when we see these people ourselves. Provided the majority of the population is doing reasonably well and those affected by an accident or bad luck need help in the short term for a rather short time. Where the majority of the population requires help and those who are doing reasonably well only represent a minority, things look different. There is a willingness to help, but anyone with common sense sees that he could give his last shirt away for the poor and still not change much. People in developing countries who are doing well themselves are no more or less helpful than the average person in wealthier countries. But their basic attitude is more realistic.

With its development away from hunting and gathering, also the mental framework and the gods and powers people believed in gradually changed. With the development of differentiated high cultures, monotheistic religions emerged, including the Jewish, Christian, Islamic,

Hindu, Buddhist beliefs, and several more. These religions competed with the older systems of belief, including animistic and older multi-goddess spiritual systems, as we know them from the pre-Christian worlds of gods in the classical Greek, Roman, Norse, Hindu, and Chinese myths,[1-5] and many more. With the expansion of European powers, the Christian belief had for a few centuries worldwide a dominant position. But not everything done in the name of the Christian religion was nice and humane.

Like anything and everyone, also the institutions run by the Christian faith said and say one thing and often did and do the other. In the course of our lives, we learn to differentiate between appearances and realities. For other things, it can take decades and even more for the dark truths to come to light. Our western society is currently working on how children of indigenous people and/or members of weaker social groups were taken away, "educated," and abused in schools, boarding schools, and other compulsory institutions and often killed or died in some other way. All this happened in the belief in the spiritual superiority of the "modern" Christian value systems and worldviews. Canada is currently being shaken by the discovery of mass child graves that symbolize the dark side of Christianization and "civilization." The children were taken away from their indigenous parents. They were forbidden to speak their mother tongue. The Truth and Reconciliation Commission of Canada rightly speaks of cultural genocide.[6-11] Single mothers, socially weak individuals, and indigenous people fared no better around the world. Now, a 100 and more years after the gloss and varnish of Christian churches has started to peel off, lawyers, historians, and other scholars around the world begin to uncover atrocities that fundamentally contradict our current system of values.[12-14]

An integral part of all large monotheistic religions is the obligation to give alms and to help the poor. But these religions all emerged in times when everyone was still responsible for himself. Fairy tales and fables tell of kings and other powerful people who helped the poor. But it was rather the exception that such help occurred. In early human history, it was about giving alms. Today's welfare in wealthy countries stands in contrast to the harsh realities of the past. The poor have a right to be helped. And what is poor? It is no longer defined by the length of their clothing, as this was the case in the European Middle Ages (we better not discuss here how often they washed these clothes and themselves), or by the fact that poor people were often on the verge of starvation. Poverty has become a sociological term in which what the individual has access to is compared with the access of other people in carefully produced statistics. If this access falls below a certain percentage of the majority, the person or the group of people he or she belongs to are considered poor.

Are we back to the beginning, where we have to distinguish what we see ourselves from what sociology tells us? Yes and no. Anyone who is qualified as "poor" in countries like the USA, Canada, Germany, Sweden, or Switzerland has a standard of living that people in many other countries would envy. That is the reason why so many people are trying to get into the USA, Canada, or Europe, despite the continuous criticism in the "progressive" press about all the injustices that exist in these countries. Those on social assistance in a well-off country often have more money available than someone who works all day in a poorer country. And so far we discussed only individual people. As soon as we look at whole countries, it gets even more complicated. Argentina was a rich country a 100 years ago, but today it cannot pay its foreign debt. Singapore was not an independent state a 100 years ago. Today it is rich and extremely densely populated. Whether countries are rich or poor today depends on what the respective leadership of the country has done and is doing, the level of social cohesion that exists in these countries, the existence of established social structures, industries, and many more factors. It is considered unfair to remember that the HIV/AIDS epidemic in South Africa with its alarmingly high infection rates is also related to the fact that the South African government denied for many years the role of the human immunodeficiency virus (HIV) and deceived its largely poor and little educated population about the real reasons for this epidemic. But that is another story.

In September 2021, the search terms "social inequality" showed over 500,000 publications in PubMed; the search terms "social inequality health" over 300,000 publications, and the search terms "social inequality COVID-19" almost 4000 publications. And on the left side-bar, we can see that the number of all papers identified with these search terms is increasing exponentially. A first conclusion is therefore that many, many authors try either to contribute to a better understanding of the pandemic and potential avenues to overcome it, to become known through such a publication, or both. Even trying to categorize the many different publications is already challenging. Some discuss issues in specific countries, such as COVID-19 in Libya, Brazil, the continent of Africa, and more. Others discuss the impact of COVID-19 on patients with a specific disease, such as HIV, sickle cell disease, gastrointestinal diseases, and more. A third group discusses the impact of COVID-19 on healthcare professionals, be it worldwide or in a specific country. It is not possible for a human being to read all these articles, not even for the challenges related to COVID-19, let alone on a general level with several hundreds of thousands of papers.

The giving of alms and the willingness to help by donations are deeply rooted in our culture. Charity is to give money, food, clothes, or time as a humanitarian act to persons in need.[15] Today, charities want to serve the public interest, common good, or philanthropy.[16] Charity aims at specific social problems, while philanthropy attempts to address more the more fundamental roots of the challenges. Charity and philanthropy are different, but they also overlap.[17] It is challenging to get an overview of the world of charities. The US-based Charity Navigator claims to be the world's largest and most trusted nonprofit evaluator.[18] It lists as categories of charities (the numbers of charities within the respective category in brackets): animals (531); arts, culture, humanities (1343); community development (926); education (714); environment (457); health (927); human services (2614); human and civil rights (398); international (642), religion (476); and research and public policy (225). In the USA, total giving to charitable organizations was $US 410.02 billion in 2017.[19] Giving has increased continuously since 1977 except for a few years. Health charities received $US 38.27 billion 9% of all donations.[20] As there is an overlap with other charity categories that relate to health challenges, the sum of health-related donations is certainly even higher. There are charities for specific diseases such as enzyme deficiencies, cystic fibrosis, multiple sclerosis, cancer, leukemia, and many more. Many more charities work worldwide beyond those in the USA. They all want to improve a lot of humanity, animals, the environment, and more.

It does not take a genius to see that poorer people have worse cards when it comes to dealing with the COVID-19 pandemic. As with so many other facets of the COVID-19 pandemic, just wielding the moral club and accusing that the world is unjust and that the pandemic has made it even more unjust will not contribute much to understanding the challenges we face.

In the following, we discuss several dimensions that overlap, but that is better examined separately for an analysis of the complex challenge we are looking at. Again, we depart from the fundaments of modern society and work our way to the effective vaccines that were developed in the Western world and are now registered by the FDA and the EMA. We discuss the value of the vaccination developed in Russia and China in Chapters 5 and 16.

2 Social framework and the responsibility for one's fate and health

In the past, most rich people were well dressed, well-fed, often overweight, often big-headed, and often arrogant. Most poor people were ill-dressed, ill-fed, intimidated, and submissive. How have times changed! Today, well-off people tend to be of normal weight, pay attention to a healthy diet, play sports, and try to strike a good balance between work and leisure. Their children go to good schools, learn from the world's problems, and often leave their homes for a year or more to help people in poorer countries. It has gone so far that parents have delivered their children to alleged orphanages because young people from rich countries wanted to help orphans, and parents assumed that their children would have it better in such an "orphanage" than in their own poor home. Dozens of websites sell volunteer trips everywhere from Uganda to China to Costa Rica—offering visitors both an opportunity to help the less fortunate, and a vacation far from the beaten path. Alleged orphanages or other types of children's homes attract voluntourists from around the world. This is part of the new and booming billion-dollar global voluntourism industry, especially popular with American teenagers taking a gap year before heading off to college.[21] In 2019, the British government updated its travel advice to discourage tourists from visiting or volunteering in orphanages. In 2020, the Dutch parliament held a debate on the practice and its connection to human trafficking. For decades, spending time in orphanages has been a popular voluntourism activity. Critics call it "orphanage tourism."[22]

In the past, being overweight was not considered a major problem. On the contrary, it was a sign of well-being. Only with the increasing longevity of the general population and since modern medicine and science recognized how harmful being overweight, lack of exercise, poor diet, and the resulting diseases can be in the long run, has our attitude gradually changed. As discussed in the previous chapters, between what we see with our own eyes, what we see on television, and about what we read, the complex worlds of modern communication have established themselves as an own level of reality, which partially distort things, and partially help us to process our immediate impressions. We know that today. We now know that obese people in affluent countries are more likely to come from the poorer classes, eat the wrong food, and go to cheap restaurants. We can see the various abuses of their bodies. But not all want to know the truth. "Fome zero" (Portuguese for *zero hunger*) was a program introduced by the Brazilian President Luiz Inácio Lula da Silva in 2003, to eradicate hunger and extreme poverty.[23] During this campaign, newspapers, journals, and television showed triumphantly how people looked at the Brazilian beaches: not undernourished, but overweight. There are people in Brazil who are hungry. But the true health challenge today is that most people eat too much, eat an unhealthy wrong composition of food, and move and exercise too little.

The USA had with Barack Obama a first black president, and has since 2021 a first female Vice President. In her inauguration speech, she emphasized how women had to fight for their place in social recognition step-by-step, and that in this respect she stands on the shoulders of many, many women in the past who have contributed to the emancipation of women.

Some diseases make life difficult. There is no justice in one person getting a rare disease and the other smoking a lifetime and staying healthy. But the vast majority of people can decide for themselves whether and what professional training they want to do, what to eat, and how much they exercise. In the countries of the West, this applies to people of every gender, every race, and every minority. We do not live in a paradise. Our western society is competitive. The best jobs are not distributed secretly, but in theory should go to those who are best suited to it in terms of their qualifications, training, and life experience. Of course, there are networks through which those who are socially adept are given preference. Again: we do not live in paradise. You need to be flexible, smart, cunning, and many more things. Life has many different aspects. One aspect is that it is a great adventure. If you are systematically prevented from personally developing in one country, you can go to another country. In fact, going abroad at a young age is always recommended. You learn things that you could not even imagine existed. This is not always easy and requires courage, energy, and perseverance. In the fairy tale worlds of soap operas on television, the reality is presented differently. Part of growing up in developed countries is also learning to look through the promises made in soap operas or through advertising. Just buying a subscription to a gym or buying expensive home training machines is not enough to get physically fit and stay fit. You have to move, you have to overcome your laziness, and you have to persevere for a long time.

3 Social inequality: The sociology approach

There is today an entire industry that thrives on the analysis and accusation of social inequality. This industry is called "sociology." In its wording: "Social inequality is an area within sociology that focuses on the distribution of goods and burdens in society."[24] Sociological observations are for example, that South Africa is home to the largest population of people living with HIV, estimated at 7.1 million as of 2016; while representing only 0.7% of the world's population, the country accounted then for 17% of the global HIV population.[25] Such observations include the moral accusation that this is unjust. There are uncountable books, written by professors in sociology and allied sciences who describe social inequality about age, gender, race, origin, sexual orientation, and many other factors, including its historical roots.[26–29] There are thousands of young persons who want to improve the world, and so they study social inequality in one of the uncountable university courses in all its nuances and facets and learn that a minority of humanity possesses more than the great majority. The persevering ones of them will write a doctoral thesis. A few ones will manage to get a university professorship in sociology. Some more may make it into a position where they can advise the government on how taxpayers' money can be distributed to support this or that disadvantaged group. The vast majority of these young person's that in their beginning are full of enthusiasm and the wish to improve the world will later work in some field that has almost nothing to do with their university training. Universities are another temptation of our modern times.

Our whole society is always a balance between different forces. To some extent, the numbers and percentage distributions published by sociologists, statisticians, and others help to get an idea. But these percentage distributions are of little help in trying to understand the forces that create these inequalities and the avenues for individuals and people collectively to make things better.

4 Ideologies and politics that promise(d) to abolish poverty

With the Enlightenment and the development of modern science and industrialization, many thought systems emerged and developed, trying to understand the rapid development of mankind, to comprehend it spiritually, and first to predict and later also to direct its future development. A set of political philosophies originated in the revolutionary movements of the 18th century out of concern for the social problems that were associated with industrialization. Before the industrial revolution, most of the workforce had been employed in agriculture. The population grew rapidly in the 19th century. With the transition from agriculture to production in industry and the growth of the cities also overcrowded slums developed. Clean water, sanitation, and public health facilities were inadequate. Literature grew up condemning the unhealthy conditions. Conditions improved over the course of the 19th century due to new public health acts regulating things such as sewage, hygiene, and home construction. The roll-out of industrialization and the move from agriculture to living in urban centers is continuing until now worldwide.[30,31]

One key aspect of the past centuries was the rise of nationalism. The European states vied to see who would become the greatest, strongest, richest, and most influential. The preliminary result of this competition was the First World War, in the wake of which millions of people were killed, a lot of wealth and culture was destroyed, several European royal families went under, and democracy was established in many countries. World War I also contributed to the drowning of several ancient and harmful ideas and structures of thought. To what extent the newly emerging ideas and thought structures were better is another story.

The desire to understand the ever faster developing world was reflected in very different intellectual approaches. "An Inquiry into the Nature and Causes of the Wealth of Nations," usually quoted by its shortened title "The Wealth of Nations," was first published by Adam Smith in 1776 at the beginning of the Industrial Revolution with the intention to describe and understand how nations build up wealth. It is today regarded as a fundamental work in classical economics. The topics it discusses include the origin and use of money, the division of labor, productivity, free markets, and more. With this book, he established the foundations of classical free-market economic theory, claiming that rational self-interest and competition lead to economic prosperity. Essentially, in Smith's understanding, the primal moving force is human nature, driven by the desire for self-betterment and guided (or misguided) by the faculties of reason.[32–35]

In opposition to the developing free-market economic theory, theoretical systems emerged that focused on the observation of poverty in the industrializing world. Karl Marx and his collaborator Friedrich Engels developed an entire philosophical, economic, and political set of ideas in opposition to the market-driven economy which they called "capitalism." They predicted that sooner or later the capitalist system would collapse and lead to a socialist revolution that would gradually lead to post-capitalist forms of society shaped by social and humane thoughts. They did not develop these thoughts in isolation but within a strong international socialist movement. They called the desired post-capitalist system socialism or communism. Many tried to develop a mindset to serve as a red line for the development of an alternative to the current societal structure that they perceived as deeply unjust. Amid these thoughts, the works of Marx and Engels offered the best systematized and most convincing theory and ideology. In the Manifesto of the Communist Party, Marx and Engels called for the political unification of the European working classes in order to achieve a communist revolution and predicted that a workers' revolution would first occur in industrialized countries. In their view, the future socio-economic organization of communism would represent a higher form of living and society than capitalism.

Other political currents did not want a radical revolution, but rather gradual improvements. The terms "social," "socialist," "social democratic," and similar terms appeared in all of these movements. In the 20th century, social democracy and communism had become two dominant political tendencies that competed with more conservative thoughts. Socialism became a very influential movement in the 20th century. Socialist parties and ideas became a political force with varying degrees of power and influence on all continents, heading national governments in many countries around the world, in various degrees of contrast to the more conservative forces.[36,37] Of course, Marx and Engels had a very low opinion of their ideological counterpart Adam Smith.

The European expansion had started shortly before 1500 and unfolded worldwide over the following centuries, caused and accompanied by the Enlightenment, industrialization, and the development of modern systems of thought. In the beginning, almost all European states were monarchies, ruled by kings who had inherited their position. The French monarchy was overthrown by a popular revolution in 1789. Other monarchies fell after World War I and II, with today's result of a hodgepodge of very different heads of state. However, in all states, step by step, universal voting right was introduced, followed eventually by the right to vote for women and people of all races. The USA declared I independence in 1776 and successfully defended it in the War of Independence, while Canada formally remained loyal to the British Crown in theory, but became independent over time.

While many movements that aimed at improving the general population's wellbeing tried to improve gradually, the idea of being able to change the entire "system" appeared to offer for a while a third avenue. In the 19th century, the terms "socialism" and "communism" were still rather interchangeable.

The Russian Vladimir Lenin developed the theory that a revolutionary vanguard party, recruited from the working class, should lead the working class (the "proletariat") to a political revolution. This would be the dictatorship of the proletariat and would help gradually first Russia and then the rest of the world to go first to socialism and from there to communism. Toward the end of the First World War, the Russian communist party, headed by Lenin, managed to take power in Russian. The former Russian ruler, tsar Nikolaus II, was later murdered on the orders of the ruling communist party along with his family and several servants, who had not belonged to the ruling class. Lenin and his party established the United Socialist Soviet Republic (USSR), whose conceptual basis was based on the teachings of Marx and Engels but further developed as a theoretical system by Lenin, who justified the dictatorship of the proletariat. The result was a dictatorship whose cruelty and inhumanity reached almost unimaginable dimensions. Lenin died rather early and was followed by Stalin, a ruthless criminal who, however, managed to give himself the appearance of a benevolent super-father and ruler. Gulag was the government agency in charge of a network of forced labor camps set up by order of Lenin, which reached its peak under Stalin 1930s to the early 1950s. The term gulag is also used to describe all forced-labor camps that existed in the USSR.[38]

In Germany, after the turmoil of World War I, Adolf Hitler came to power, whose ideology mixed components of German nationalism with socialist ideas, resulting in the German "National Socialism" (Nationalsozialismus), with the national socialists (the "Nazis") ruling. The result was a different form of dictatorship, as grueling and cruel as that of the Soviet Union. The competition between these two gentlemen and the other European states seeking supremacy eventually led to World War II, in which the US-allied with England, and ultimately won. Japan had allied with Germany, had occupied large parts of Asia, and was finally defeated by the US and its allies. Large parts of China had been occupied by Japan. In the end, the Chinese Communist Party drove the Japanese out. Under Mao Zedong, it established a rule that was based on the teachings of Marx, Engels, Lenin, and Stalin. The Chinese communist party is in power until today. In a few more countries communist parties came to power, establishing paradises on earth in North Korea, Cuba, Zimbabwe, Albania, and a few more.

The theory of socialism sounds nice and humane but contradicts human nature. Man can also be altruistic, but in the long run, he does not live on promises, but on the fact that his needs are met, that he has a job, a house, a family, and is left in peace by the state. None of the socialist states could offer this in the long run. For a while, many of them were able to suppress the people with violence and terror. The Soviet Union imploded and was dissolved in 1991. Most former members of the USSR became independent states. Russia became the main successor to USSR, renamed itself the Russian Federation, assumed the USSR's permanent membership in the United Nations (UN) Security Council, and inherited the USSR's entire nuclear arsenal. Russia is no longer ruled by the Communist Party. Instead, it has become an "ordinary" dictatorship, at present headed by a former member of the secret service. While Russia has stalled economically, China instead allowed after a while the forces of the market and is now one of the largest economic powers of the world. It is still run by the Communist Party and follows, in theory, the teachings of Marx, Engels, Lenin, and Stalin. A schizophrenic challenge for anybody who wants to make a career in China. Whatever you say (or think), be careful not to be viewed as critical of the prevailing order. You better do not say anything that might displease the Communist Party, but as the Chinese economy is running, you also should not express revolutionary demands for equal distribution of wealth.

The world has changed economically, technically, scientifically, socially, and in all its thought structures since the beginning of industrialization half a millennium ago. Today, we are in a new phase of the industrial revolution with computer-based technology rapidly developing in increasingly more areas. The industry that developed and produced drugs during the Second World War was initially the chemical industry, then became the pharmaceutical industry, then the life science industry, and is now in transition to the biopharmaceutical industry.

All radical prophecies that promised to abolish poverty resulted in dictatorships, horrifying murders, and wars with millions of deaths. Scientifically, the two main outcomes of World War II were the development of the atomic bomb and the industrialized production of penicillin.

When mankind is confronted with new, confusing situations, people seek consolation and guidance in the thoughts that already exist, but originated on a different background that no longer exists. This includes religions, belonging to a specific ethnic group, and also the socialist thoughts that have always promised everything possible and impossible, and have always achieved the opposite.

Few of those see themselves today as socialists, social democrats, or the like dream of a radical socialist revolution. The historical experiences of the socialist revolutions in Russia and other countries were simply too repulsive for that, especially today since we have access to the archives of that time and the atrocities that occurred are well documented. There are also no longer individual theorists who can claim to be respected worldwide as fundamental thinkers of socialist thought. "Social" and "socialist" thinking has developed together with sociology into an intellectual mish-mash that tries to reduce social inequality through excessive state intervention. Of course, you need a lot of money for that. This is then done by increasing taxes.

5 Developing countries and the COVID-10 pandemic

It was, to a relevant extent, the availability of modern drugs that contributed to the extension of longevity in all parts of the world. However, we are also seeing a transition toward making higher demands on the part of the developing countries. The World Health Organization (WHO) model list of essential medicines contains the medications considered to be most effective and safe to meet the most important needs in a health system. It is divided into core items and complementary items. The core items are seen as the most cost-effective options, while the complementary items either require additional infrastructures such as specially trained health care providers or diagnostic equipment or have a lower cost–benefit ratio. While the initial list contained predominantly generic drugs, the WHO demands patented drugs have increased.[39]

The history of mankind was not a democratic process where people could vote to which empire they wanted to belong. History was and is cruel and full of wars. The process of colonization was brutal. It was initially based on the superiority of the ships of the European states and their cannons.[40] However, superior ships and weapons technology are not sufficient to explain the colonization of the world by European states. Further factors were the desire to Christianize, the pursuit of own material interests by the involved individuals, including openly criminal energy, brutal assertiveness, and a myriad of other factors. After World War II, the colonial powers were forced to retreat between 1945 and 1975, when nearly all colonies gained independence.[41]

It was the life science industry and biotechnology companies that we're able to quickly develop effective vaccines against the COVID-19 pandemic. But it was not the richest countries that tackled vaccination coverage the fastest. In many rich countries, governments have embarrassed themselves to the bone with fickle decisions, late ordering of emerging vaccines, and inconsistent communication regarding the need for vaccinations.

Many countries that were initially able to hold back the spread of the COVID-19 pandemic by isolating and reducing contact with infected people, then hesitated too long to consistently vaccinate the population. If the wealthy countries had quickly ordered the vaccines in sufficient quantities and organized the vaccination of their populations, then the COVID-19 pandemic would already be controlled much better today than is currently the case.

The scheme currently used by the WHO to differentiate the world into low-to-middle income countries and high-income countries does not do justice to reality. From the WHO perspective, the richer countries are sitting on the vaccines, while the poorer countries need them at least as badly. For months, the Director-General of the WHO has spoken out against boosting vaccinations with a third booster shot. Instead, has called on wealthy countries "with large supplies of coronavirus vaccines" to refrain from offering booster shots through the end of the year, expanding an earlier request that had been largely ignored. Speaking to journalists WHO Director-General Tedros Adhanom Ghebreyesus also said he was appalled at comments by a leading association of pharmaceutical manufacturers saying that vaccine supplies with high enough to allow for both booster shots and vaccinations in countries in dire need of jabs but facing shortages. In his opinion, we should not accept countries that have already used most of the global supply of vaccines, using even more of it, while the world's most vulnerable people remain unprotected.[42–44]

These remarks will not advance the global fight against the COVID-19 pandemic. They are rhetoric that sees the world in the logic of a zero-sum game, which is a situation in which an advantage that is won by one of two sides is lost by the other. If the total gains of the participants are added up, and the total losses are subtracted, they will sum to zero. Cutting a cake, where taking a more significant piece reduces the amount of cake available for others is a zero-sum game if all participants value each unit of cake equally.[45] Specifically in the view of a worldwide pandemic, a view that sees the world as static is a temptation that misses the bottom line. Such rhetoric works in a bazaar negotiation for the best price. But the fight against the COVID-19 pandemic is not a bazaar negotiation.

It is a new high-tech challenge for mankind that requires the most modern technology to solve. In fact, this technology does not yet exist in every country on earth. The reasons for this are complex. Past injustices should not be talked away here. The big question is not only how the people of the developing countries can be vaccinated as quickly as possible, but how it happens amid a strategy that opens the avenue to an industrialized and technology-based economy for these countries or promotes this opening. For a long time, the governments of the well-off countries did not understand the fundamental importance of a resolute vaccination of the population. They overslept during the valuable time. Should these governments follow the WHO's demand and refrain from a third booster vaccination? Booster vaccinations further limit the risk of becoming infected, especially for vulnerable people. The mistrust in governments is rightly low in the well-off countries too. Refraining from booster vaccinations, perhaps to help people in other countries is the opposite of what democratically elected governments in free countries should do. They would not survive the next election.

Furthermore, we have not only observed reasonable government decisions made by developing countries' governments during the COVID-19 pandemic. Brazil's President is trying successfully to appear even more openly idiotic than his great role model Donald Trump. The difference is that behind the US President stood the machinery of the biotech and life science industry, which could at least get him to initiate Operation Warp Speed (OWS). Brazil does not have such potential. Tanzania proudly claimed to have remained COVID-19 free. In the end, the president died of COVID-19, although of course that was not officially admitted.[46]

6 The misconception of a weakening of intellectual property as a way out of the pandemic

Intellectual property is a category of property that includes intangible creations of the human intellect. It is a distinct type of property compared to physical property like gold, land, or houses. The most well-known types are copyrights, patents, and trademarks, but there are several more. Only in the late 20th century, the concept of intellectual property became commonplace, in the majority of the world's legal systems. A patent gives its owner the right to exclude others from making, using, or selling an invention for a limited period, in exchange for sharing details of the inventions with the public. Incentives of the patent system include encouragement to invent at all; to disclose the made invention; to invest the funds necessary to experiment, produce and market the invention; and to design around and improve upon earlier patents.[47,48]

Patents and other forms of intellectual property protection have played an essential role in encouraging the development first of drugs and in the last decades also of biopharmaceuticals. The process of developing a new drug and bringing it to the market is long, costly, and risky. However, once the respective drug is on the market, it is relatively easy and cheap to produce a generic version. The involved patents block the entry of cheap generics into the market for a limited period. For companies to be motivated to continue to invest in innovative development, they must be able to expect that they can charge enough during this period to recoup costs and make a profit. After patents expire, generic competition will enter the market. Generic companies do not have high development costs, sell at a lower price, and can and will thereby erode the innovator drug company's revenues.[49–52]

The huge spread between the development costs for drugs and the comparatively low production costs are used for decades by opponents of the pharmaceutical industry to defame it as greedy for money. The argument is neither new nor particularly reasonable but is of course also used by the advocates for the curtailment of intellectual property rights in favor of developing countries.[25,53] We saw what the development potential of the life science and biopharmaceuticals industry is worth during the COVID-19 pandemic. Without them, we would not have any effective vaccines and would have to rely on questionable products from Russian, Chinese, or Cuban production.

Patents are the lifeblood of any pharmaceutical, biotech, or life science company. Without their protection, there would be no future for the respective company.[54] In 1998, 39 pharmaceutical manufacturers sued the government of South Africa to prevent the implementation of a law that intended to facilitate access to HIV/AIDS drugs at low cost. The companies accused South Africa, the country with then the largest population of individuals living with HIV/AIDS worldwide, of circumventing patent protections guaranteed by intellectual property rules that were included in the latest round of world trade agreements. The companies dropped their lawsuit in the spring of 2001 after an avalanche of negative publicity.[55] Since then, companies have become more aware of the potential damage that can be caused by a too strict interpretation of intellectual property rights. Working in collaboration with national governments, organizations such as the WHO, and non-governmental organizations such as the Bill and Melinda Gates Foundation and pharmaceuticals companies have begun to explore avenues through the minefield of intellectual property protection in less developed countries. Most now have donation schemes for drugs to treat diseases such as leprosy and HIV.[54]

The Agreement on Trade-Related Aspects of Intellectual Property Rights (TRIPS) is an international legal agreement between all the member nations of the World Trade Organization (WTO). It establishes minimum standards for the regulation by national governments of different forms of intellectual property. In 2001, the Doha Declaration declared that TRIPS should not prevent states from dealing with public health crises and allowed for compulsory licenses.[56] Since the 1990s, activists have continued to demand a suspension of intellectual property rights for drugs that are able to treat life-threatening diseases, claiming that patients and governments in developing countries cannot afford modern effective drugs because they are too expensive.

It is to no big surprise that now a growing chorus of advocates wants to weaken the intellectual property protections for vaccines against COVID-19.[57] Activists and academics that want to support developing countries demand that their governments should prepare to issue compulsory licenses of vaccines and effective COVID-19 treatments. Compulsory licensing is allowed by the TRIPS agreement and should enable governments to supply their citizens with generic versions of patented treatments either through domestic production or imports.[53] Such demands are further supported by statistical/sociological statements such as that high-income countries represent only 16% of the world's population, but have purchased more than half of all COVID-19 vaccine doses.[58]

There is no world government, and there is no right for the numerical majority of the world's population to impose their will on the rest of humanity. The numerical data on the percentage of people who have a majority of existing doses of vaccines are superficial and misleading. Before one can argue about who should receive what share of the existing vaccines, these vaccines must first exist. Their development includes planning; the implementation and financing of the necessary clinical studies, and the acceptance and execution of additional regulatory requirements that are usually meaningful, but unfortunately sometimes pointless as well. Then comes the production of these vaccines, their storage, distribution, and finally their administration by qualified personnel. The proposals made by activists would require that the manufacturing capacity to manufacture sufficient quantities of vaccines were available also in developing countries; as well as capacity to store them at very low temperates; to distribute them without interrupting the chain of refrigerated storage; and much more.

We have seen in the previous chapters how the prophets of wealth distribution promise to take everything away from the rich to improve the world in one swoop. Unfortunately, the reality is much more complicated. It is one thing to make radical demands to the general public that cost nothing. Who would dare to speak out against poor people being vaccinated against COVID-19 as soon as possible?

The development of effective vaccines against COVID-19 was a huge step forward in the development of vaccination technology and in the record speed at which all of the required clinical trials were conducted. Only a few very advanced biotech companies were able to achieve this technical and logistical masterpiece. Companies have to survive in economic competition. Humanistically inspired advocates of the interests of less affluent countries tend to forget this. If the patent rights of companies for the COVID-19 vaccines were curtailed, they would no longer be interested in working with the same energy in the next pandemic.

In May 2021, under pressure from activists and Democratic lawmakers, the administration of US President Joe Biden reversed course and announced its support for a vaccine intellectual property waiver. Proponents say the waiver could boost the production of vaccines and other life-saving products. TRIPS commits members to guarantee a minimum level of intellectual property protection. But it was the intellectual property rule, and the profits they allow, that facilitated the development of the advanced affected COVID-19 vaccines. Low manufacturing capacity, not patents, are at present the biggest impediment to global vaccination efforts.[59,60] Taking away intellectual property protection would not be a good step forward to resolve the challenges of the COVID-19 pandemic.

7 Conclusions

It was a remarkable achievement how quickly effective COVID-19 vaccines were developed and approved. Without these vaccines, modern human society would be severely restricted in its further development. It is now a matter of time before international trade, tourism, the exchange of goods, and the rest of the economy grow again. It is right and legitimate that the developed countries initially thought of their people above all else. We are now at a new stage in the fight against this epidemic. Except for really medically justified non-vaccinations, the restrictions for those unwilling to vaccinate should be tightened. Vaccination of the people of the developing world should now be pushed ahead with all vigor but without a moral club.

References

1. Wikipedia. *Greek Mythology*. https://en.wikipedia.org/wiki/Greek_mythology.
2. Wikipedia. *Roman Mythology*. https://en.wikipedia.org/wiki/Roman_mythology.
3. Wikipedia. *Norse Mythology*. https://en.wikipedia.org/wiki/Norse_mythology.
4. Wikipedia. *Hindu Mythology*. https://en.wikipedia.org/wiki/Hindu_mythology.
5. Wikipedia. *Chinese Mythology*. https://en.wikipedia.org/wiki/Chinese_mythology.
6. National Center for Truth and Reconciliation. https://nctr.ca/records/reports/.
7. Truth and Reconsiliations Commission of Canada. https://www.rcaanc-cirnac.gc.ca/eng/1450124405592/1529106060525.
8. Report of the Truth and Reconciliation Commission of Canada; 2015. http://caid.ca/DTRC.html.
9. Riley NS. *Perspective: Thousands of Children Died in Canadian Schools. We're Just Learning Where They're Buried*. Deseret News; August 7, 2021. https://www.deseret.com/u-s-world/2021/8/7/22610323/canada-unmakarked-graves-thousands-of-children-died-in-residential-schools.
10. Honderich H. *Why Canada is Mourning the Deaths of Hundreds of Children*. Washington, DC: BBC News; July 15, 2021. https://www.bbc.com/news/world-us-canada-57325653.
11. Wikipedia. *Canadian Indian Residential School Gravesites*. https://en.wikipedia.org/wiki/Canadian_Indian_residential_school_gravesites.
12. Daly K. Conceptualising responses to institutional abuse of children. *Curr Issues Crim Just*. 2014;26(1):6–29. http://www.austlii.edu.au/au/journals/CICrimJust/2014/10.pdf.
13. Sköld J. The truth about abuse? A comparative approach to inquiry narratives on historical institutional child abuse. *Hist Educ*. 2016;45(4):492–509. https://doi.org/10.1080/0046760X.2016.1177607.
14. Mäkelä D. *Historical Child Abuse in out-of-Home Care: Finland Disclosing and Discussing its Past (Master's Thesis in Child Studies)*. Linköping University; May, 2015. http://liu.diva-portal.org/smash/get/diva2:844350/FULLTEXT01.pdf.
15. Wikipedia. *Charity (Practice)*. https://en.wikipedia.org/wiki/Charity_(practice).
16. Wikipedia. *Charitable Organization*. https://en.wikipedia.org/wiki/Charitable_organization.
17. Wikipedia. *Philanthropy*. https://en.wikipedia.org/wiki/Philanthropy.
18. *Charity Navigator—About Us*. https://www.charitynavigator.org/index.cfm?bay=content.view&cpid=8658.

References

19. *Charity Navigator*. https://www.charitynavigator.org/index.cfm?bay=search.categories&categoryid=5.
20. *Charity Navigator—Giving Statistics*. https://www.charitynavigator.org/index.cfm?bay=content.view&cpid=42.
21. Moran T, Jesko J. *The Dark Side of Orphanage 'Voluntourism' in Nepal That's Putting Children at Risk. Experts warn that donations are sometimes going to orphanage owners' pockets*. ABC News; August 9, 2016. https://abcnews.go.com/International/dark-side-orphanage-voluntourism-nepal-putting-children-risk/story?id=41239651.
22. Lu J. *Why There's a Global Outcry Over Volunteering at Orphanages*. NPR, Goats and Soda; January 13, 2020. https://www.npr.org/sections/goatsandsoda/2020/01/13/779528039/why-theres-a-global-campaign-to-stop-volunteers-from-visiting-orphanages?t=1631008969269.
23. Wikipedia. *Fome Zero*. https://en.wikipedia.org/wiki/Fome_Zero.
24. *Social Inequality. Social Inequality is an Area Within Sociology that Focuses on the Distribution of Goods and Burdens in Society*. University of Oslo, Department of Sociology and Human Geography; 2019. https://www.sv.uio.no/iss/english/research/research-areas/social-inequality/.
25. 't Hoen EFM, Kujinga T, Boulet P. Patent challenges in the procurement and supply of generic new essential medicines and lessons from HIV in the southern African development community (SADC) region. *J Pharm Policy Pract*. 2018;11:31. https://www.ncbi.nlm.nih.gov/pmc/articles/PMC6277991/pdf/40545_2018_Article_157.pdf.
26. Warwick-Booth L. *Social Inequality*. 2nd ed. Thousand Oaks, CA: SAGE Publishers; 2018.
27. Sernau SA. *Social Inequality in a Global Age*. 6th ed. Thousand Oaks, CA: SAGE Publishers; 2019.
28. Carr D. *Golden Years?: Social Inequality in Later Life*. New York, NY: Russell Sage Foundation; 2019.
29. Savage M. *The Return of Inequality: Social Change and the Weight of the Past*. Cambridge, MS: Harvard University Press; 2021.
30. Wikipedia. *Industrial Revolution*. https://en.wikipedia.org/wiki/Industrial_Revolution.
31. Wikipedia. *Industrialisation*. https://en.wikipedia.org/wiki/Industrialisation.
32. Wikipedia. *The Wealth of Nations*. https://en.wikipedia.org/wiki/The_Wealth_of_Nations.
33. Wikipedia. *Adam Smith*. https://en.wikipedia.org/wiki/Adam_Smith.
34. Smith A. *The Wealth of Nations*. North Charleston, SC: CreateSpace Publishers; 2014.
35. Britannica. *Adam Smith*. https://www.britannica.com/biography/Adam-Smith.
36. Wikipedia. *Socialism*. https://en.wikipedia.org/wiki/Socialism.
37. Wikipedia. *Marxism*. https://en.wikipedia.org/wiki/Marxism.
38. Wikipedia. *Gulag*. https://en.wikipedia.org/wiki/Gulag.
39. *WHO Model List of Essential Medicines*. https://en.wikipedia.org/wiki/WHO_Model_List_of_Essential_Medicines.
40. Cipolla CM. *Guns, Sails, and Empires: Technological Innovation and the Early Phases of European Expansion, 1400–1700*. New York, NY: Barnes Noble Books; 1996.
41. Wikipedia. *Colonialism*. https://en.wikipedia.org/wiki/Colonialism.
42. *WHO Urges Rich Countries to Hold Off on Booster Shots Until 2022. Agency Chief Calls for an Extension of Proposed Moratorium on COVID-19 Vaccine Booster Shots to Enable Every Country to Inoculate At Least 40 Percent of Its Population*. Aljazeera; September 8, 2021. https://www.aljazeera.com/news/2021/9/8/who-chief-urges-halt-to-booster-shots-for-rest-of-the-year.
43. WHO. *Director-General's Remarks at the Press Conference for the Inauguration of the WHO Hub for Pandemic and Epidemic Intelligence*; September 1, 2021. https://www.who.int/director-general/speeches/detail/director-general-s-remarks-at-the-press-conference-for-the-inauguration-of-the-who-hub-for-pandemic-and-epidemic-intelligence.
44. Beaubien J. *Why WHO Is Calling for a Moratorium on COVID Vaccine Boosters*. NPR; August 4, 2021. https://www.npr.org/sections/goatsandsoda/2021/08/04/1019780576/why-who-is-calling-for-a-moratorium-on-covid-vaccine-boosters?t=1631251621589.

45. Wikipedia. *Zero-sum game*. https://en.wikipedia.org/wiki/Zero-sum_game.
46. Magufuli J. *Tanzania's President Dies Aged 61 After Covid Rumours*. BBC News; March 18, 2021. https://www.bbc.com/news/world-africa-56437852.
47. Wikipedia. *Intellectual Property*. https://en.wikipedia.org/wiki/Intellectual_property.
48. Wikipedia. *Patent*. https://en.wikipedia.org/wiki/Patent.
49. Grabowski HG, DiMasi JA, Long G. The roles of patents and research and development incentives in biopharmaceutical innovation. *Health Aff (Millwood)*. 2015;34(2):302–310. https://www.healthaffairs.org/doi/pdf/10.1377/hlthaff.2014.1047.
50. Barton JH, Emanuel EJ. The patents-based pharmaceutical development process: rationale, problems, and potential reforms. *JAMA*. 2005;294(16):2075–2082.
51. Atkinson JD, Jones R. Intellectual property and its role in the pharmaceutical industry. *Future Med Chem*. 2009;1(9). https://www.future-science.com/doi/pdf/10.4155/fmc.09.138.
52. Kesik-Brodacka M. Progress in biopharmaceutical development. *Biotechnol Appl Biochem*. 2018;65(3):306–322. https://www.ncbi.nlm.nih.gov/pmc/articles/PMC6749944/pdf/BAB-65-306.pdf.
53. Wong H. The case for compulsory licensing during COVID-19. *J Glob Health*. 2020;10(1), 010358. https://www.ncbi.nlm.nih.gov/pmc/articles/PMC7242884/pdf/jogh-10-010358.pdf.
54. Yarza C. *Intellectual Property Protection*. PriceWaterhousCooper (pwc); 2021. https://www.pwc.com/il/en/pharmaceuticals/intellectual-property-protection.html.
55. Barnard D. In the high court of South Africa, case no. 4138/98: the global politics of access to low-cost AIDS drugs in poor countries. *Kennedy Inst Ethics J*. 2002;12(2):159–174.
56. Wikipedia. *TRIPS Agreement*. https://en.wikipedia.org/wiki/TRIPS_Agreement.
57. Owens C. *The Growing Fight Over Coronavirus Vaccine Patents*. Axios; March 30, 2021. https://www.axios.com/coronavirusvaccine-patents-biden-55431035-fb17-4eed-8c76-e9f0ade04b09.html.
58. Sheikh AB, Pal S, Javed N, et al. COVID-19 vaccination in developing nations: challenges and opportunities for innovation. *Infect Dis Rep*. 2021;13(2):429–436. https://www.ncbi.nlm.nih.gov/pmc/articles/PMC8162348/pdf/idr-13-00041.pdf.
59. Siripurapu A. *The Debate Over a Patent Waiver for COVID-19 Vaccines: What to Know*. Council on Foreign Relations (CFR); May 26, 2021. https://www.cfr.org/in-brief/debate-over-patent-waiver-covid-19-vaccines-what-know.
60. Nicolás ES. *EU Counters Biden's Vaccine Patent-Waiver With WTO Plan*. EU Observer; June 4, 2021. https://euobserver.com/coronavirus/152056.

CHAPTER 14

Politics, illusions, websites, and the real world

1 Humanity and communication

Everybody lives at the same time in different worlds. There is the physical and social world, immediately around us. The smaller the community in which we live, the more manageable it is, and, depending on the basic attitude, the more comfortable or the more boring. The latter attitude is certainly the privilege of persons in puberty and the years thereafter. However, at the same time, we all are part of many more communities. We hear to the radio, we look at the television, we read newspapers and books, and today most people walk around with an iPhone or a comparable gadget.

Only a tiny minority of people still live as hunters and gatherers. Most people in industrialized countries now carry an identity (ID) card with them or at least a driver's license where there are no IDs. We live in states with laws that we more or less comply with. Most of us can read and write and have access to the internet and other tools of the cyber-sphere. We know that there is a big, wide world beyond our immediate world of life. Some of us travel a lot for work and speak several languages while others feel at home in their home environment and only leave it now and then to go on vacation or to visit friends. The opening of the entire world to the individual took place in several steps in the distant past. They are difficult to categorize, but there is no doubt that the speed of globalization is increasing.

Humans have evolved away from their animal origins on two levels. Physically through the development of the upright gait, through the differentiation between hands and feet, the loss of massive body hair, and many more details. In parallel and in interaction with it, human communication developed with intellectual constructs that went beyond the individual. In early societies, the physical world, the spiritual world, and the belief that the whole world and all of its constituents have a soul were not yet as rationally separated as today, where there is established scientific knowledge and we are able to differentiate to a certain extent between imagination and reality. In the early spiritual world, objects, places, animals, the wind, the rain, and all other elements of nature and mankind had their own spiritual essence, including love, peace, and war.[1-4] The emergence of this early interpersonal spiritual world would not have been possible without human language. Additional elements are non-verbal communication, music, dance, physical interaction, smelling, and more, dimensions that we partly share with other higher developed animals. But it was the language that played the key role in the development of interpersonal cultural worlds. To be able to speak, the vocal organs had to change their properties to allow human speech, including the development of the tongue, the lips, the vocal cords, the larynx, the oral cavity, and more, from the animal beginnings to today's human organs that are coordinated by the central nervous system, allowing speech, singing, snoring, whistling, and more.[5,6] How the interaction and the feedback between body, spoken language, and the emerging spiritual and cultural worlds developed will continue to keep busy armies of scientists for the next thousands of years after we have at least already begun to span a wide range in the understanding of human development without ideological blinkers.[7,8]

The next key step was the emergence of writing, which occurred independently in several high cultures around the world several 1000 years ago. In Europe, we inherited the Latin letters we are currently using from the ancient Romans, who developed them using older models from the other Mediterranean and Middle East cultures.[4] The numerals we are currently using are called Arabic numerals, but they were invented in India, were adopted by Arabs, spread to medieval Europe during the Middle Ages, and were exported worldwide together with the Latin alphabet with the European expansion.[9-11] Arabic numerals are even used in Chinese. The writing was first used to document purchases, such as 10 goats or five sacks of grain. In the beginning, the writing systems were complicated, only a small group of specially trained people mastered the art of reading and writing. During the centuries and millennia that followed, the writing systems became simpler. The writing was no longer just used to document economic transactions, but to write down legal texts, poems, literature, religious texts, and more.

Literature and the spiritual worlds in which people lived were never limited to the written texts that played initially a relatively small role on the fringes of human society. Oral transmission of fairy tales, legends, stories, and historical contexts played a major role. They were now gradually supplemented by written texts. All aspects of human communication developed in mutual interaction. With advancing technology and science, not only have the technical possibilities expanded, but the spheres of the communicated worlds have expanded and partially developed a life of their own.

One turning point was the development of letterpress printing at the end of the middle ages in Europe. Books had previously been copied by hand and were very precious. Now they became affordable. Another key element was the invention of paper, another international story. Papermaking was invented in China, spread from there to the Islamic world, and reached from there Europe in the late Middle Ages, as trade began to grow again, literature and art began to flourish, and science and enlightenment began to take their course. Parchment made from animal hides had been the predominant writing material. Now it became too slow and expensive to make in the fast of the new fast-growing need. The European worldwide expansion was not only accompanied by sailing ships and guns,[12] but also by documents printed books, and records of all kinds on paper. Produced originally from rags, plants, and mulberry bark, the papermaking process advanced further with the industrialization until the invention of wood-based paper.[13,14] There are many more things to report in the unfolding of human literacy, art, culture, technology, and civilization. But these are other stories.

The internationalization of trade, the development of the colonial empires, access to new spices from the Far East, and all the things that made life around us more colorful have placed the world around us more and more in the context of influential and diverse international currents. Key elements like metals, letters, and numerals developed internationally, but it took centuries or even millennia for them to spread. With the acceleration of technology development, everything became faster.

2 Fairy tales, oral tradition, the worlds of radio and television, and the internet

New technologies allowed communication with people far away, including telegraphy, radio, landline telephone, radio, television, and finally mobile phones that enable us to see and communicate with distant family members, colleagues, business partners, and more in their own country and increasingly also in remote provinces and countries. The next big step came with computers, the internet, and today's emerging artificial intelligence technology. Shortly, we are sure to face further key steps.

In the Middle Ages and before, people had learned and talked about the worlds of the rich and powerful through conversation, gossip, fairy tales, songs, and other oral channels. Printed news and the introduction of radio and television established new worlds that most ordinary people had previously only known from hearsay. Increasing school education also played an essential role. Somehow, of course, everyone knew that not everything that was portrayed on the radio or television was real. But these media allowed new dimensions of additional dreams and ideas of worlds that had previously been almost completely closed to ordinary citizens.

Nothing fundamentally essential has changed. We still differentiate between what we believe in personally and what other people, institutions, doctrines, religions, or more tell us. We have different levels of trust and skepticism towards the various sources of information that reach us. Fairy tales were mostly told in the evening and everyone knew that most of it were fantasy. Although there were also expectations, real hopes, and truths hidden in them, inexperienced people found the worlds of television reports, documentaries, and soap operas to be convincing at first.

The internet gives us access to information that in the past used to require physical visits to the library. The internet has ruined time-honored institutions like the Encyclopedia Britannica because no one bought the many heavy tomes anymore. Today also this institution predominantly publishes on the Internet.[15] Wikipedia as an internet-based encyclopedia has secured a dominant place. Whatever we remember from past readings, we can refresh it with a few mouse clicks.

No one can read everything on the internet, no one can look at all the movies, videos, and more, or look at all the texts and pictures distributed through social media and all the other information and misinformation circulating in the cyber-sphere. Every government, institution, company, and increasingly also many private individuals have their website. Most of them show their best side with their internet presence. How much of the impression they try to make correspond to the reality and is correct is another question. Again, nothing of all this is fundamentally new. You learn to read sideways, and you block what you think is junk. You sleep over the news; discuss it with people you trust, sometimes with people you distrust. Eventually, you come to your own opinion, and then make your personal decisions - what to wear, to buy, to sell, whether you want to meet face to face someone you came in contact through the internet, tinder, or one of the uncountable other matchmaking services; and thousands of others decisions.

3 Politics, websites, and the real world

Scientific publications endeavor to describe and analyze the world and its myriad of facets rationally, to build knowledge that goes beyond the individual expressions of opinion of individual researchers. Good scientific journals do not publish every nonsense, but let other peer scientists check whether the statements made are understandable and contribute to the development of more knowledge.

But the spheres in which we live also include feelings, music, premonitions, and much more. Before communication technology, music could only be heard when someone was physically playing an instrument in front of the audience. Theater allowed actors to be observed, but only as long as actors and spectators were physically nearby. All of that has changed radically with communication technology. The World Wide Web is a global information medium that anybody can access via a computer connected to the internet. It was created in 1990 by a physicist working for the European Organization for Nuclear Research (CERN—the acronym that represents the organization's name in French), the largest particle physics laboratory in the world. Its early origins go further back to the transfer of documents and files from one server. It developed fast into the global system of interconnected computer networks as we know it today. It has enabled individuals and organizations to publish ideas and information to a potentially large audience online at greatly reduced expense and time delay.

In the roughly 30 years of its existence, the internet has revolutionized worldwide communication, with an increasing percentage of all human beings participating. Already more than half of the world's population has today access to the internet.[16]

The internet and the technologies that create cyberspace are transforming society, business, and politics. In the online environment the combination of technology, information, and instinctive mental processes contribute to reshaping how people think.[17]

Every government, governmental institution, university, scientific association, scientific journal, and many more have today their website. These websites serve several purposes. They are intended to provide information, but they also represent the respective institution as well as possible. When we look at the websites of the European Medicines Agency (EMA) or the World Health Organization (WHO), we see wonderful, politically correct portrays of how these institutions work tirelessly for the good of humanity.[18,19] Their actors may even believe in it themselves, but of course, that is only one side of the story.

4 The COVID-19 pandemic and the internet

In its beginning, the COVID-19 pandemic has often been likened to a war-like state against an invisible enemy in which healthcare workers were described as frontline soldiers, overwhelmed by patient numbers, ventilator shortages, and rationing of personal protective equipment. In the beginning, there was indeed relatively little known about the virus, its clinical course, treatment strategies, or how to organize the healthcare systems to meet this challenge.[20]

During the COVID-19 pandemic, the use of the internet and social media has virtually exploded. As with all things in life, there are positive and negative things to say here.[21,22] Updates on current situations were quick and easy, but the internet was also flooded with false reports, misinformation, and all sorts of nonsense. Ever since the first headline reported the COVID-19 outbreak, social media has served as a breeding ground for the contagion of information about the novel coronavirus. The information, a potpourri of true observations, misunderstandings, fears, courageous reports, and lies, has exploded across various social media platforms, outpacing the spread of the disease. A social media pandemic accompanies and follows the disease pandemic, stirring a wide spectrum of emotions. The world has witnessed pandemics before, but all were in the pre-social media era. The assembling of information vs misinformation, trust-building vs fear-mongering, and anger vs comfort are a few of the sentiments reverberating in the social media pandemic. In the new reality of social distancing and self-quarantined lockdown, Twitter has emerged as a paramount platform for crisis communication.[23]

An "infodemic" emerged with false news, conspiracy theories, magical cures, and racist news that were shared at an alarming rate, with the potential to increase anxiety and stress and even lead to loss of life. Also, journals and authors published a mass of academic articles at a speed that suggests a non-existent or a non-rigorous peer-review process.[24] Billion people in lockdown had easy access to information owing to easy and affordable internet connectivity and electronic media. But information overload and misinformation during the pandemic also posed a set of challenges that were not encountered before. A couple died because they took excess doses of chloroquine sulfate in Arizona. More than a 100 Iranians died from alcohol poisoning because of the belief that industrial-grade ethanol would protect them against the virus infection. There are many more such examples.[25,26] Furthermore, social media posts with misleading or dangerous opinions and analyses were often amplified by celebrities and social media influencers, contributing substantially to an avalanche of misinformation. A counterbalance to this harmful infodemic were physicians and scientists who saw a public presence as a large segment of their mission and who tried to bring authority and real-world experience to the COVID-19 discussion.[27] However, a potential challenge in all the messages brought to the public by the specialists who were suddenly constantly present on radio, television and other media was that otherwise, few people listen to what they have to say. Now suddenly the media were interested in their opinion. This went hand in hand with the risk that these specialists would market themselves. Some of them massively overestimated the relevance of their own opinion, continued with grave warnings, demanded harsher lockdown rules, and more. As always, a certain pinch of caution and mistrust is the order of the day.

The rise of the COVID-19 pandemic has changed individuals' lives in both positive and negative ways. Access to the application of various digital infrastructures is increasingly becoming considered to be essential. The rapid application of the new technological methods to curb the pandemic has resulted in a worldwide broad digital division. Some states were well prepared, some not at all. In developed countries, policies such as lockdowns and social distancing measures have resulted in technological advancement and new means of interaction with government, businesses, and citizens, including increased online shopping, robotic delivery systems, digital and contactless payment systems, remote working, distance learning, telehealth, 3D printing, online entertainment, and more. Such technological advancements have been embraced during this pandemic by the technically advanced countries around the world.[28] Also the advancements in artificial intelligence techniques played a major role in biomedical sciences attempting to handle the various aspects of the COVID-19 pandemic.[29] Things look different in less developed countries. Some governments ignored the pandemic, others copied blindly the approach of developed countries. But they ignored that people who do not have a permanent job and have to get by daily cannot afford to simply not to go to work. For them, there is no work in the home office. The COVID-19 pandemic has accentuated the inequality in our world. But there is nothing new about that, and neither has the internet been able to change that.

5 Conclusions

It is always the same. On a government level, the world was rather unprepared. Lessons had not been learned from previous epidemics and pandemics.[30] People always draw from many sources. They have their physical environment; their social environment; their cyber-world environment; their inner world, which processes all these outer worlds permanently, their experiences; their instincts; and their gut feelings. Individuals navigate between these elements with different levels of trust and mistrust and end up making their personal decisions. And also institutions and governments navigate. The more one has learned to differentiate between appearance and reality, between honesty and lies, between inner desires and the most harsh reality, the better this navigation process will go.

References

1. Harvey G. *Animism—Respectivng the Living World*. Columbia, New York, NY: Columbia University Press; 2005.
2. Wikipedia. *Animism*. https://en.wikipedia.org/wiki/Animism.
3. Wikipedia. *Communication*. https://en.wikipedia.org/wiki/Communication.
4. Wikipedia. *History of Communication*. https://en.wikipedia.org/wiki/History_of_communication.
5. Crelin ES. *The Human Vocal Tract: Anatomy, Function, Development, and Evolution*. New York, NY: Vantage Press; 1987.
6. Bermejo-Fenoll A, Panchón-Ruíz A, Sánchez Del Campo F. Homo sapiens, chimpanzees and the enigma of language. *Front Neurosci*. 2019;13:558. https://www.ncbi.nlm.nih.gov/pmc/articles/PMC6555268/pdf/fnins-13-00558.pdf.

7. Harari YN. *Sapiens—A Brief History of Humankind*. New York, NY: Harper; 2015.
8. Diamond J. *Guns, Germs, and Steel: The Fates of Human Societies*. New York, NY: Norton; 1991.
9. Robinson A. *The Story of Writing: Alphabets, Hieroglyphs & Pictograms*. 2nd ed. London, UK: Thames & Hudson; 2007.
10. Wikipedia. *History of Writing*. https://en.wikipedia.org/wiki/History_of_writing.
11. Wikipedia. *Hindu-Arabic Numeral System*. https://en.wikipedia.org/wiki/Hindu%E2%80%93Arabic_numeral_system.
12. Cipolla CM. *Guns, Sails, and Empires: Technological Innovation and the Early Phases of European Expansion, 1400–1700*. New York, NY: Barnes Noble Books; 1996.
13. Kulansky M. *Paper: Paging through History*. New York, NY: W. W. Norton & Company; 2017.
14. Wikipedia. *History of Paper*. https://en.wikipedia.org/wiki/History_of_paper.
15. *Encyclopedia Britannica*. https://www.britannica.com/.
16. Britannica. *Internet*. https://www.britannica.com/technology/Internet.
17. Lewis JA. *A Short Discussion of the Internet's Effect on Politics*. Center for Strategic & International Studies (CSIS); January 29, 2021. https://www.csis.org/analysis/short-discussion-internets-effect-politics.
18. European Medicines Agency (EMA). www.ema.europa.eu.
19. World Health Organisation (WHO). www.who.int.
20. Aggarwal NR, Alasnag M, Mamas MA. Social media in the era of COVID-19. *Open Heart*. 2020;7(2), e001352. https://www.ncbi.nlm.nih.gov/pmc/articles/PMC7526301/pdf/openhrt-2020-001352.pdf.
21. Venegas-Vera AV, Colbert GB, Lerma EV. Positive and negative impact of social media in the COVID-19 era. *Rev Cardiovasc Med*. 2020;21(4):561–564. https://rcm.imrpress.com/EN/10.31083/j.rcm.2020.04.195.
22. Bin Naeem S, Kamel Boulos MN. COVID-19 misinformation online and health literacy: a brief overview. *Int J Environ Res Public Health*. 2021;18(15):8091. https://www.ncbi.nlm.nih.gov/pmc/articles/PMC8345771/pdf/ijerph-18-08091.pdf.
23. Kumar A, Khan SU, Kalra A. COVID-19 pandemic: a sentiment analysis. *Eur Heart J*. 2020;41(39):3782–3783. https://www.ncbi.nlm.nih.gov/pmc/articles/PMC7454503/pdf/ehaa597.pdf.
24. Mheidly N, Fares J. Leveraging media and health communication strategies to overcome the COVID-19 infodemic. *J Public Health Policy*. 2020;41(4):410–420. https://www.ncbi.nlm.nih.gov/pmc/articles/PMC7441141/pdf/41271_2020_Article_247.pdf.
25. Sasidharan S, Harpreet Singh D, Vijay S, et al. COVID-19: Pan(info)demic. *Turk J Anaesthesiol Reanim*. 2020;48(6):438–442. https://www.ncbi.nlm.nih.gov/pmc/articles/PMC7720829/pdf/tjar-48-6-438.pdf.
26. Tsao SF, Chen H, Tisseverasinghe T, et al. What social media told us in the time of COVID-19: a scoping review. *Lancet Digit Health*. 2021;3(3):e175–e194. https://www.ncbi.nlm.nih.gov/pmc/articles/PMC7906737/pdf/main.pdf.
27. Topf JM, Williams PN. COVID-19, social media, and the role of the public physician. *Blood Purif*. 2021;50(4–5):595–601. https://www.ncbi.nlm.nih.gov/pmc/articles/PMC7900472/pdf/bpu-0001.pdf.
28. Renu N. Technological advancement in the era of COVID-19. *SAGE Open Med*. 2021;9, 20503121211000912. https://www.ncbi.nlm.nih.gov/pmc/articles/PMC7958161/pdf/10.1177_20503121211000912.pdf.
29. Rasheed J, Jamil A, Hameed AA, et al. COVID-19 in the age of artificial intelligence: a comprehensive review. *Interdiscip Sci*. 2021;13(2):153–175. https://www.ncbi.nlm.nih.gov/pmc/articles/PMC8060789/pdf/12539_2021_Article_431.pdf.
30. Radanliev P, De Roure D, Walton R, et al. COVID-19 what have we learned? The rise of social machines and connected devices in pandemic management following the concepts of predictive, preventive and personalized medicine. *EPMA J*. 2020;11(3):311–332. https://www.ncbi.nlm.nih.gov/pmc/articles/PMC7391030/pdf/13167_2020_Article_218.pdf.

CHAPTER 15

Global warming, Armageddon warnings, and the COVID-19 pandemic

1 Climate change and global warming: The basics

The term "global warming" became very popular after North America Space Agency (NASA) climate scientist James Hansen had used it in a testimony to the US Senate on climate change in 1988. Since then, a broad awareness of global warming emerged. The terms "global warming" and "climate change" are often used interchangeably. Both imply massive human responsibility. There is widespread scientific agreement that we are seeing gradual warming of the earth in our decades. There have been previous periods of climatic change in the past millennia of the earth's history, but in our century were observed an unprecedented impact of mankind's activities on the earth's climate system. The largest driver of warming is the emission of gases that create a greenhouse effect, of which more than 90% are carbon dioxide (CO_2) and methane. The burning of fossil fuels, including peat, wood, coal, oil, and natural gas to provide energy is the main source of these emissions, with additional contributions from agriculture, deforestation, and manufacturing. Temperature rise is accelerated or tempered by climate feedbacks, such as loss of sunlight-reflecting snow and ice cover, increased water vapor (a greenhouse gas itself), and changes to land and ocean carbon sinks. A further key player is aerosols, suspensions of fine solid particles, or liquid droplets in the air. Aerosols can be natural or man-made. Fog, mist, or geyser steam are natural aerosols. Particulate air pollutants and smoke are man-made. Aerosols in the lower layer of the atmosphere have an atmospheric lifetime of only about a week, while aerosols in the higher layers of the atmosphere (the stratosphere) can remain there for years. Aerosols on one side limit global warming by reflecting sunlight, but black carbon in soot that falls on snow or ice can also increase global warming by increasing the absorption of sunlight. Man-made aerosols have been declining since 1990 and thus are no longer a counterbalance against greenhouse gas warming. Evidence of warming from air temperature measurements is supplemented by a wide range of other observations. There has been an increase in the frequency and intensity of heavy precipitation, melting of snow and land ice, and increased atmospheric humidity. Flora and fauna are also behaving in a manner consistent with warming. Plants are flowering earlier in spring. Also, the cooling of the upper atmosphere shows that greenhouse gases are trapping heat near the Earth's surface and prevent it from radiating into space.[1-3]

Climate change and global warming have local, national and international sides. Worldwide, there is an increased density of populations; in the USA an explosive growth of human activity, including low-density rural housing. In the USA and other countries, wildfire suppression has contributed to the accumulation of fuel loads in the forests, increasing the risk of large, catastrophic fires. In the USA, this wildfire suppression was contrary to traditional land management methods practiced by indigenous peoples. Limited wildfires had contributed to cleaning the forest of bio-debris and fuels. With bio-debris and fuels accumulating, mega-fires have a much stronger destructive power. Today, most wildfires are human-caused, although lightning remains a common cause, too. Climate changes have provided more heat, but also seem to increase the number and frequency of lightning strikes. If your house burns down due to a wildfire, you are this very moment less interested in reflecting global causes. But already this example shows the involved complexity. Wildfires are not just a consequence of global warming. Other factors are the existence of working forest administrations, professional and well-equipped fire brigades, well-maintained fire-fighting aircraft for large-scale operations, or the lack of all the institutions just listed. In some countries, forests have been reforested with trees that grow quickly. That looks nice and green. But no one thought that such trees also burn quickly. In addition, there are diametrically opposed interests. In many southern European countries, fires are set to remove forest because the culprits assume that they will then be able to build houses on the new treeless land. In the Amazon, ranchers have an interest in burning down forests and preventing re-forestation in order to have more grazing land for their cattle. The world has become complex, and at the state level a good balance of the various interests is desirable. But there is corruption, laziness, narrow-mindedness, and incompetence in some or many state administrations. The discussion about how to prevent forest fires and wildfires or put them out if necessary has become a science in itself both on the local, the national, and the international level.[4-6]

The national and later international discussion of climate change and global warming began in specialized academic circles and societies in the 19th century. It expanded in the 20th century to the general international political institutions of the world and has developed a lively interaction between these institutional levels. The Intergovernmental Panel on Climate Change (IPCC) was established in 1988 by the World Meteorological Organization (WMO) and the United Nations Environment Program (UNEP),[7–9] with subsequent endorsement by the UN General Assembly also in 1988. The World Meteorological Organization (WMO) is a specialized UN agency promoting international cooperation on atmospheric science, climatology, hydrology, and geophysics. It originated from the International Meteorological Organization, a nongovernmental organization founded in 1873 as a forum for exchanging weather data and research.[10] The prehistory of the International Meteorological Organization goes even back one or two more centuries.[11] In contrast, the United Nations Environment Program (UNEP) is relatively young and was established in 1972. The initial task of the Intergovernmental Panel on Climate Change (IPCC) was to prepare a comprehensive review and recommendations for the state of knowledge of the science of climate change; the social and economic impact of climate change, and potential response strategies and elements for inclusion in a possible future international convention on climate. The IPCC does not conduct original research and does not monitor climate changes. Instead, it undertakes systematic reviews of relevant literature to provide a comprehensive update on climate change, its effects, and potential strategies. Worldwide, scientists and other experts review and surmise the data and compile key findings into assessment reports. Since 1988, the IPCC has delivered five such assessment reports. It has also produced a range of methodology reports, special reports, and technical papers, in response to requests for information on specific scientific and technical matters from the United Nations Framework Convention on Climate Change (UNFCCC, see below), governments and international organizations.[12,13]

In 1992, the United Nations Framework Convention on Climate Change (UNFCCC) was established, aiming at stabilizing atmospheric composition to prevent dangerous anthropogenic interference with the climate system and achieve that in ways that do not disrupt the global economy.[14] The US was the first developed country that sign the UNFCCC. In the following, it was ratified by almost all countries. Defining the level of warming that would constitute "dangerous anthropogenic interference" became a crucial and demanding part of the global discussion.[15]

In 1997, the Kyoto Protocol extended the UNFCCC convention and included legally binding commitments for most developed countries to limit their emissions. During the Kyoto Protocol negotiations, the group "G77," representing developing countries,[16] pushed for a mandate requiring developed countries to take the lead in reducing their emissions, since developed countries contributed most to the accumulation of greenhouse gases in the atmosphere. In 2009, the Copenhagen Accord endorsed the

Kyoto protocol. However, it was widely portrayed as disappointing because of its low goals and was rejected by poorer countries including the G77.[1] Finally, in 2015 all UN countries negotiated the Paris Agreement. The Paris Agreement's long-term temperature goal is to keep the rise in mean global temperature to well below 2 °C above pre-industrial levels, and preferably limit the increase to 1.5 °C, recognizing that this would substantially reduce the impacts of climate change. Emissions should be reduced as soon as possible and reach net-zero in the second half of the 21st century.[17,18]

In 2019, the UK parliament declared officially a climate emergency as the first national government worldwide, followed by other countries. In November 2019 the EU Parliament declared a "climate and environmental emergency,"[19] and the European Commission presented its European Green Deal intending to make the EU carbon-neutral by 2050. The promised benefits of the European Green Deal include fresh air, clean water, healthy soil, and biodiversity; renovated, energy-efficient buildings; healthy and affordable food; more public transport; cleaner energy and cutting-edge clean technological innovation; longer-lasting products that can be repaired, recycled and re-used; future-proof jobs and skills training for the transition; and globally competitive and resilient industry.[20,21] Major countries in Asia have made similar pledges: South Korea and Japan have committed to becoming carbon neutral by 2050, and China by 2060.

A quick reminder of the EU Lisbon Agenda, devised in 2000 as an action and development plan devised for the economy of the EU between 2000 and 2010. It proclaimed the plan to make Europe by 2010 "the most competitive and the most dynamic knowledge-based economy in the world."[22] However, this has not occurred. The EU has fallen back in the international competition. Advances in space travel (SpaceX, Blue Origin, Sierra Nevada Corporation, Boeing),[23] the lodging business (Airbnb),[24,25] the personal transportation business (Uber),[26,27] the introduction of electric cars (Tesla),[28] and many other areas came from US initiatives. Some of them were copied in China. The EU is watching and sometimes adopting elements. At least electric cars are now produced worldwide.[29] The EU Lisbon Agenda was succeeded by the Europe 2020 strategy with even more flowery promises.[30] Governmental pronouncements, international protocols, and EU agendas usually sound promising at first glance. But as already discussed in the previous chapters, they not only contain concrete plans, but also flowery promises, soap-box speeches, and self-portraits. Another challenge of the 21st century. It is becoming more and more important to maintain common sense and not to trust everything that bears the stamp of international science and institutions. We must maintain the same critical eye when looking at international plans to control climate change.

As climate change and global warming are in public discussion now since decades, a lot of diverging positions have emerged. Apart from the usual textbooks,[3,31,32] some books emphasize that the fight against climate change and global warming has evolved into a mega-billion dollar industry with its self-interests.[33] Another book explains the broad worldwide agreement among scientists with "group think," i.e., the pressure of the majority.[34] In fact, a scientist who can see positive aspects of global warming, for example, that wine is now also grown in countries in which this was not possible in the past due to their past cold climate, will have much worse prospects for an academic career than other young scientists who howl with the wolves and complain and denounce the extinction of bat species, beetle species, and other creatures. Another author warns that mankind might destroy its species and thereby foreclose the potential of a living universe.[35]

2 The predicted effect of climate change on humans

The predicted effects of climate change on humans include forecasted effects on health, environment, displacement, migration, security, society, human settlement, energy, and transport. Climate change is claimed to have brought about possibly irreversible alterations to Earth's geological, biological, and ecological systems, resulting in large-scale environmental hazards to human health; extreme weather, the higher danger of wildfires, loss of biodiversity, stresses to food-producing systems, and the global spread of infectious diseases. Furthermore, most adverse effects ascribed to climate change are reported for poor and low-income communities around the world, with much higher levels of vulnerability and much lower levels of capacity available for coping with environmental change.[36]

3 Global warming and the COVID-19 pandemic

The search terms "climate change COVID-19" resulted in almost 700 publications in PubMed in October 2021. Many authors claim a direct causal relationship between ongoing climate change, the current COVID-19 pandemic, and other pandemics in the recent past. Several authors speak of two current pandemics, COVID-19 and climate change.[37,38] Others speak of threats to human existence from climate change, biodiversity loss, and a pandemic that is devastating economies and paralyzing societies,[39] or climate change, air pollution, and COVID-19.[40] Another author claims that the pandemic arrived at a time when wildfires, high temperatures, floods, and storms amplified human suffering.[41]

Some authors are using the pandemic to outline a global path to a better future for all of humanity. Most of these publications see a key role for scientists in overcoming the COVID-19 pandemic and for a path for humanity to a better future. One author claims that the most striking similarities between COVID-19 and climate change are the consequences of ignoring science and a corollary belief in the authority of one's personal beliefs. However, he sees it as a good thing about science that it allows people to understand the world and how it works. In his view, the COVID-19 pandemic has delivered an uncomfortable reminder of the value of trusting in science. Therefore, we should look toward science and scientists concerning the climate crisis.[42] Effective global surveillance, global political will, and multidisciplinary collaboration of all stakeholders are promoted to help respond to any future threats.[43] Yet another author mentions that the spread of diseases has been associated with air pollution and social inequities, such as racial discrimination, gender inequality, and racial, economic, and educational disparities. Instead, there should be a future where we are equal in all respects, where no one threatens us, where we can live healthy and happy lives. For this future, all obstacles should be removed.[44]

The first Nobel Prize summit "Our Planet, Our Future" was an online meeting to discuss the state of the planet at a critical juncture for humanity. It brought together Nobel Laureates and other leading scientists with thought leaders, policymakers, business leaders, and young people to explore solutions to immediate challenges facing our global civilization: mitigate and adapt to the threat posed by climate change and biodiversity loss, reduce inequalities and lift people out of poverty; all this is now seen as even more urgent due to the economic hardships posed by the pandemic; science, technology, and innovation should enable societal transformations while anticipating and reducing potential harms. The authors see an urgent need for people, peoples, economies, societies, and cultures to actively start governing nature's contributions to wellbeing and building a resilient biosphere for future generations and to reconnect development to the Earth system foundation through active stewardship of human actions into prosperous futures within planetary boundaries.[45] Yet another author sees COVID-19 as a potential catalyst to set the global society on a new path to a sustainable relationship between people and the rest of nature: a greening of human society. Referring to the book "Half-Earth, Our Planet's Fight for Life,"[46,47] he recommends devoting half of Planet Earth's land to environmentally sound management, with the top 25% of the land is legally protected areas managed by national or provincial conservation agencies.[41]

The COVID-19 pandemic is characterized not as a crisis concerning health, but as a crisis about life itself, revealing the hubris of the view that assumes the omnipotence of mankind and seeing human activity as the dominant influence on the future of life on our planet.[15] Others construct a direct causal connection between the increased diversity of bats, global warming, and the emergence of the COVID-19 pandemic.[48] Some emphasize the similarities between the two crises, even suggesting that climate change may have been a causal factor in the COVID-19 pandemic. They outline that, although the COVID-19 pandemic and climate change do not immediately appear similar, upon closer inspection several significant shared factors can be seen. Both crises are attributed to substantial unnecessary loss of life.[49,50] A growing number of representatives from geoscience, epidemiology, computational intelligence and many more are calling for the findings of their respective specialist disciplines to be included in the overall debate about global warming and its alleged consequences.

There are several overarching key elements and assumptions in these publications. Both COVID-19 pandemic and global warming are perceived as catastrophes that have caused even more harm in our world and continue to do so than was already there. In this view, science provided the essential elements to overcome the pandemic. Science will help to overcome air pollution and social inequities, such as racial discrimination, gender inequality, and racial, economic, and educational disparities, or similar goals dressed in other words.[38,40,41,44,51,52] Therefore, science now needs to be supported even more in order to make even further progress. And, as to be expected, the well-known demands are repeated that money should be transferred from the military budget to the health research and public health budget.[39,53]

The European Health Emergency Preparedness and Response Authority (HERA) aims at setting the priorities for an EU environment, climate, and health research agenda "by adopting a holistic and systemic approach in the face of global environmental changes."[54] HERA is now working on its Health and Environment Research Agenda 2020–2030,[55] setting priorities in research on environment, climate, and health, and to identify research needs regarding COVID-19.[56,57] HERA has established six work packages (WPs), listed in an abbreviated form in the following[58]:

- WP1: Ensures effective and efficient coordination and management of the HERA consortium.
- WP2: Develop a cross-cutting stakeholder community in Environmental Health and produce the 2020–2030 European environment, Climate, and Health research agenda. Collaborates with relevant stakeholder communities and networks at the National, Regional, and European levels.
- WP3: Identify proactively key areas in policy and practice that require scientific support in the environment, climate, and health nexus in Europe in the next decade and to structure the work in an inclusive

way ensuring recognition of all relevant regional, national, and EU perspectives of stakeholders including those from European countries with the less developed environment and health research and policy.
- WP4: Uncover research gaps and needs in the area of environment, health, and climate. Integrates the knowledge and forecasts further development and needs. WP discusses health effects of environmental exposures; global population increase and urbanization, industrialization, and geopolitical problems that have accelerated global changes altering the nature and impact of the environment on health; recent efforts to mitigate the threats to health and ecosystems; and more. The focus will be on the WHO definition of health "State of complete physical, mental, and social well-being, and not merely the absence of disease or infirmity"; this definition of health comes from the WHO declaration of Alma Ata of 1978,[59] see also Chapters 8 and 11).
- WP5: Move toward a harmonized methodology in health impact assessment studies considering environmental factors and the societal and economic impacts thereof and develop new ones integrating broader factors.
- WP6: Provide support to other work packages and communicate and disseminate information and knowledge with/to various stakeholder groups and communities; develop resources for effective communication and dissemination of HERA results. Ensure that the impact of HERA is maximized through effective communication to existing coordinating bodies/platforms or newly established national/regional hubs within the Environment and Health (EH) processes, agreements, communities, and networks and initiatives. Disseminate the results, including outreach/educational activities targeting different audiences to build/strengthen their capacities, with special attention paid to countries/communities with low-intensity EH research.[58]

Together with 35 individual authors, the HERA-COVID-19 working group has published a paper on the research needs emerging with the COVID-19 pandemic and global environmental change. In this paper, they summarize that what started as a health crisis has become a "social, economic, environmental and political challenge," for which it is imperative that recovery plans and taxpayers' funds go beyond the current economic and social crisis, support and reinforce the EU's ambitious Green Deal and Sustainable Development Goals (SDGs) to benefit health and efficiently prevent future crises.[54]

The EU Green Deal is a set of policy initiatives by the European Commission with the overarching aim of making Europe climate neutral in 2050.[20,21] There are no specific EU Sustainable Development Goals (SDGs). Instead, the SDGs refer to the 17 UN SDGs that are discussed in-depth in Chapter 13.[60,61]

Overall, we can already state here and now that the result of HERA and its many ambitious goals will be an exponential number of scientific publications that will not move or change anything in the world for themselves. But for the researchers who work in the HERA network, this represents a heavenly bonanza. Many more academic careers will flourish.

4 The mixing of geoscience, social, and medical challenges

Many authors emphasize that people who are in a vulnerable social position anyway are most affected by the COVID-19 pandemic.[15,37,62,63] There is nothing revolutionary new about this observation. A family with a solid income is better prepared for possible strokes of fate than a family that lives from hand to mouth and has no economic reserves. It is more complex with larger entities and states, but it is comparable. States, where the public administration cares little or nothing and basically leaves the citizens to their own fate, have fared worse than the states with a functioning infrastructure, a well-organized system of public health, medical care, and enough hospital beds for emergencies. Again, it is more complicated when we look at the individual states. The fate of individual poor people with COVID-19 in poor countries is heartbreaking. Just like the fate of people who became infected in well-off states at the beginning of the pandemic and died lonely in hospital or helpless at home. Nonetheless, for a rational analysis, we cannot pretend that there were and are no reasons that people in different states coped with the situation differently. We must not let ourselves be blinded by focusing only on the world's social inequality before, during, and after the pandemic.

5 Scientific warnings in the past

Since the beginning of industrialization and the Enlightenment, there have been many urgent scientific warnings about the future. In the following, just three examples.

In 1798, Thomas Robert Malthus claimed that the increase in a nation's food production improved short-term the well-being of the population, but that this was temporary because it would lead to population growth, which in turn would restore the low production level per capita. The human population, he argued, tended to increase geometrically, outstripping the production of food, which increased arithmetically. The force of a rapidly growing population against a limited amount of land meant diminishing returns to labor. The result, he claimed, was chronically low wages, which prevented the standard of living for most of the population from rising above the subsistence level. The lower class would suffer hardship, want, and higher susceptibility to famine and disease. He wrote in opposition to the popular view in 18th century Europe that saw society as improving and becoming more stable. Instead, he perceived population growth as inevitable whenever conditions improved. "The power of population is indefinitely greater than the power in the earth to produce subsistence for man."[64–66] Comparable warnings of worldwide famine have accompanied our civilization during the last centuries.

The term "forest dieback" (in German: "Waldsterben") refers to a stand of trees losing health and dying without an obvious cause. It can affect individual species of trees, but also multiple species, and may take on many locations and shapes. It can be along the perimeter, at specific elevations, or dispersed throughout the forest ecosystem, visible by falling off of leaves and needles, discoloration of leaves and needles, thinning of the crowns of trees, dead stands of trees of a certain age, and changes in the roots of the trees. The public discussion about "forest dieback" and "Waldsterben" (the German term was often used even in the English media) focused on acid rain as the cause. Possible other causes such as incorrect tillering or temporary drought were not discussed for a long time. Of course, climate change and global warming were discussed as major contributors. The United Nations (UN) Food and Agriculture Organization (FAO) published a worldwide overview of "decline and dieback of trees and forests" in 1994.[67] The public discussion gave the impression that the world's forests were going to die out completely. Political measures were taken that resulted in a significant reduction in emissions. In retrospect, we have to note that the discussion for a long time was alarmist and exaggerated. To this day, Europe has not become deforested, as was indicated in the horror scenarios.

The Club of Rome was founded in 1968 in Rome, Italy, consisting of 100 full members selected from current and former heads of state and government, UN administrators, high-level politicians, and government officials, diplomats, scientists, economists, and business leaders from around the globe. Since 2008, it is based in Switzerland. It stimulated considerable public attention in 1972 with its first report "The Limits to Growth." Based on computer simulations, it suggested that economic growth could not continue indefinitely because of resource depletion. The report went on to sell 30 million copies in more than 30 languages, making it the best-selling environmental book in history.[68,69]

In 1991, the Club of Rome published "The First Global Revolution." It analyzed the problems of humanity, calling these the "problematique." It claimed that in the past, social or political unity had commonly been motivated by enemies. Every state has been so used to classifying its neighbors as friend or foe, that the sudden absence of traditional adversaries has left governments and public opinion with a great void to fill. New enemies have to be identified, new strategies imagined, and new weapons devised.

"In searching for a common enemy against whom we can unite, we came up with the idea that pollution, the threat of global warming, water shortages, famine, and the like, would fit the bill. In their totality and their interactions, these phenomena do constitute a common threat that must be confronted by everyone together. But in designating these dangers as the enemy, we fall into the trap, which we have already warned readers about, namely mistaking symptoms for causes. All these dangers are caused by human intervention in natural processes, and it is only through changed attitudes and behavior that they can be overcome. The real enemy then is humanity itself."[68,70]

In 2001 the Club of Rome established a think tank, called *tt30*, consisting of about 30 men and women, ages 25–35, aiming to identify and solve problems in the world, from the perspective of youth. As of 2017, there have been 43 reports to the club, peer-reviewed studies commissioned by the executive committee, or suggested by a member or group of members, or by outside individuals and institutions. The most recent is "Come On! Capitalism, Short-termism, Population and the Destruction of the Planet."[71] In 2016, the club initiated a new youth project called "Reclaim Economics" to support students, activists, intellectuals, artists, video-makers, teachers, professors, and others to "shift the teaching of economics away from the mathematical pseudo-science it has become." In 2019, the Club of Rome issued an official statement in support of Greta Thunberg and the school strikes for climate, urging governments across the world to respond to this call for action and cut global carbon emissions.[68,72–74]

The statements, claims, and reports by the Club of Rome are based on pessimistic key assumptions that reflect an overall Malthusian bias. Already in 1973 an interdisciplinary team at Sussex University's Science Policy Research Unit reviewed the structure and assumptions of used models, which do not accurately reflect reality, and published its finding in Models of Doom.[68,75]

The three theoretical systems presented, i.e., the culture-pessimistic works of Malthus, the alarmistic theories about the alleged global death of forests, and the warning calls of the Club of Rome are only examples of a tradition that runs through the history of industrialization. Enlightenment, scientific development, and industrialization were not decreed from above by a government resolution. The dynamism of this development is based on the initiative of the many thousands of protagonists who each found a way in their time to think about small, medium-sized, or fundamentally groundbreaking innovations and then put them into practice. Personal initiative is essential. But of course, it needs a framework that allows it to unfold. Researchers like Craig Venter, who helped terminate the Human Genome Project against the will of most of its collaborating institutes years before it was due to close, were burned or otherwise silenced in centuries past. Pessimistic interpretations of the history of industrialization use the method that they extrapolate problems and mis-developments, which undoubtedly have always existed and will always exist in large quantities, from their present state into the future, and then claim that the world is doomed to end. Our earth does not suffer from too little food being produced. Sometimes too much is produced, is thrown away, or rots somewhere. The global famines that Malthus prophesied did not materialize. The European countries are still green and forested, partially as a consequence of the measures taken against acid rain and pollution. Despite the warnings of the Club of Rome, the planet did not go under and it will not go under anytime soon. Science, industry, and the complex modern industrialized society are good at finding solutions to challenges.

6 The privileges of youth

Young people have the privilege of seeing things more clearly than adults who have been familiar with reality for decades and have also got used to things that really should be changed and/or improved. Young people have the disadvantage that they still have little experience in life and cannot imagine that things will gradually be improved and corrected over time. This is nicely expressed in the frequently quoted saying "If you are not a liberal when you are young, you have no heart, and if you are not a conservative when old, you have no brain."[76] The great social and socialist movements of the 19th and 20th centuries led to unimaginable human catastrophes where they achieved unlimited power, cynically eclipsing everything that young people had initially fought against. The Club of Rome and the youth movement "Fridays for future" have the illusion that it will be able to save the earth quickly and thoroughly if governments only proceed radically enough. But the roots of social inequality are deep and complex.

7 COVID-19 is a high-tech challenge

COVID-19 is a high-tech challenge that has met our advanced civilization. It has killed many elderly people and many younger ones too. But without modern civilization not so many people would live, and certainly not many of the elderly and very old people. The COVID-19 pandemic is a good example of how mankind is going to overcome this challenge.

The key to overcoming the pandemic are effective vaccinations. Their development would have been impossible without the advances made in biotechnological research over the past few decades. But it would also have been impossible without the existence of a healthy pharmaceutical and life science industry, and without the balanced working relationship between the western licensing authorities and the industry. Achieving this balance has taken many decades of a fierce conflict, and these conflicts will never cease. States like Russia and China, which lack the culture of debate that is needed for such a balance, have developed vaccines that have been approved not only in their own country: two Chinese ones also by the WHO; and all of them in several developing countries (see also Chapter 5). But their effectiveness is doubtful, and those who want to be vaccinated safely will choose one of the vaccines approved by the FDA, EMA, MHRA, or PMDA. We discuss Russia and China in the next chapter.

References

1. Wikipedia. *Climate Change.* https://en.wikipedia.org/wiki/Climate_change.
2. Hansen J. Can we defuse the global warming time bomb? *Nat Sci.* 2003;1:1–16. https://pubs.giss.nasa.gov/docs/2003/2003_Hansen_ha07900q.pdf.
3. Houghton J. *Global Warming: The Complete Briefing.* 5th ed. Cambridge, UK: Cambridge University Press; 2015. 2015.
4. Ferguson G. *Land on Fire: The New Reality of Wildfire in the West.* Portland, OR: Timber Press; 2016.
5. Flannigan MD, Amiro BD, Logan KA, et al. Forest fires and climate change in the 21st century. *Mitig Adapt Strateg Glob Chang.* 2005. https://web.archive.org/web/20090325095123/https://www.firelab.utoronto.ca/pubs/2005_flannigan_wotton_etal.pdf.
6. Wikipedia. *Wildfire.* https://en.wikipedia.org/wiki/Wildfire.
7. World Meteorological Organisation (WMO). https://public.wmo.int/en.
8. United Nations Environment Programme (UNEP). https://www.unep.org/.
9. Wikipedia. United Nations Environment Programme (UNEP). https://en.wikipedia.org/wiki/United_Nations_Environment_Programme.
10. Wikipedia. World Meteorological Organisation. https://en.wikipedia.org/wiki/World_Meteorological_Organization.
11. Wikipedia. International_Meteorological_Organization (IMO). https://en.wikipedia.org/wiki/International_Meteorological_Organization.
12. United Nations Intergovernmental Panel on Climate Change (IPCC). https://www.ipcc.ch/.
13. Wikipedia. Intergovernmental Panel on Climate Change (IPCC). https://en.wikipedia.org/wiki/Intergovernmental_Panel_on_Climate_Change.
14. United Nations Framework Convention on Climate Change (UNFCCC). https://unfccc.int/.
15. Horton R. Offline: a global health crisis? No, something far worse. *Lancet.* 2020;395(10234):1410. https://www.ncbi.nlm.nih.gov/pmc/articles/PMC7252047/pdf/main.pdf.
16. Wikipedia. *Group of 77 (G77).* https://en.wikipedia.org/wiki/Group_of_77.
17. The Paris Agreement. https://unfccc.int/process-and-meetings/the-paris-agreement/the-paris-agreement.
18. Wikipedia. *Paris Agreement.* https://en.wikipedia.org/wiki/Paris_Agreement.
19. *The European Parliament declares climate emergency.* EU Parliament News; November 29, 2019. https://www.europarl.europa.eu/news/en/press-room/20191121IPR67110/the-european-parliament-declares-climate-emergency.
20. European Commission. *A European Green Deal. Striving to be the First Climate-Neutral Continent.* https://ec.europa.eu/info/strategy/priorities-2019-2024/european-green-deal_en.
21. Wikipedia. *European Green Deal.* https://en.wikipedia.org/wiki/European_Green_Deal.
22. Wikipedia. *Lisbon Strategy.* https://en.wikipedia.org/wiki/Lisbon_Strategy.

23. Wikipedia. *List of Private Spaceflight Companies.* https://en.wikipedia.org/wiki/List_of_private_spaceflight_companies.
24. Airbnb. www.airbnb.com.
25. Wikipedia. *Airbnb.* https://en.wikipedia.org/wiki/Airbnb.
26. Uber. www.uber.com.
27. Wikipedia. *Uber.* https://en.wikipedia.org/wiki/Uber.
28. Tesla. www.tesla.com.
29. Wikipedia. *List of Electric Cars Currently Available.* https://en.wikipedia.org/wiki/List_of_electric_cars_currently_available.
30. Wikipedia. *Europe 2020 Strategy.* https://en.wikipedia.org/wiki/Europe_2020.
31. Wiles J. Global Warming Handbook. 2nd ed. *Global Warming and Climate Change;* 2020. Independently Published.
32. Bill Gates. *How to Avoid a Climate Disaster: The Solutions We Have and the Breakthroughs We Need.* New York, NY: Allen Lane; 2021.
33. Sangster MJ. *The Real Inconvenient Truth: It's Warming: But It's Not CO_2: The Case for Human-Caused Global Warming and Climate Change Is Based on Lies, Deceit, and Manipulation.* Independently Published, USA; 2018.
34. Booker C. *Global Warming: A Case Study in Groupthink.* London, UK: The Global Warming Policy Foundation; 2018.
35. Rees M. *Our Final Hour: A Scientist's Warning. How Terror, Error and Environmental Disaster Threaten humankind's Future in This Century-on Earth and Beyond.* New York, NY: Basic Books; 2004.
36. Wikipedia. *Effects of Climate Change on Humans.* https://en.wikipedia.org/wiki/Effects_of_climate_change_on_humans.
37. Salas RN, Shultz JM, Solomon CG. The climate crisis and Covid-19—a major threat to the pandemic response. *N Engl J Med.* 2020;383(11), e70.
38. Joshi M, Caceres J, Ko S, et al. Unprecedented: the toxic synergism of Covid-19 and climate change. *Curr Opin Pulm Med.* 2021;27(2):66–72. https://www.ncbi.nlm.nih.gov/pmc/articles/PMC7924924/pdf/copme-27-66.pdf.
39. Garcia D. Redirect military budgets to tackle climate change and pandemics. *Nature.* 2020;584(7822):521–523. https://media.nature.com/original/magazine-assets/d41586-020-02460-9/d41586-020-02460-9.pdf.
40. Marazziti D, Cianconi P, Mucci F, et al. Climate change, environment pollution, COVID-19 pandemic and mental health. *Sci Total Environ.* 2021;(773), 145182. https://www.ncbi.nlm.nih.gov/pmc/articles/PMC7825818/pdf/main.pdf.
41. McNeely JA. Nature and COVID-19: the pandemic, the environment, and the way ahead. *Ambio.* 2021;50(4):767–781. https://www.ncbi.nlm.nih.gov/pmc/articles/PMC7811389/pdf/13280_2020_Article_1447.pdf.
42. Perkins KM, Munguia N, Ellenbecker M, et al. COVID-19 pandemic lessons to facilitate future engagement in the global climate crisis. *J Clean Prod.* 2021;(290), 125178. https://www.ncbi.nlm.nih.gov/pmc/articles/PMC7670902/pdf/main.pdf.
43. Sabin NS, Calliope AS, Simpson SV, et al. Implications of human activities for (re)emerging infectious diseases, including COVID-19. *J Physiol Anthropol.* 2020;39(1):29. https://www.ncbi.nlm.nih.gov/pmc/articles/PMC7517057/pdf/40101_2020_Article_239.pdf.
44. Hashimoto S, Hikichi M, Maruoka S, et al. Our future: experiencing the coronavirus disease 2019 (COVID-19) outbreak and pandemic. *Respir Investig.* 2021;59(2):169–179. https://www.ncbi.nlm.nih.gov/pmc/articles/PMC7832026/pdf/main.pdf.
45. Folke C, Polasky S, Rockström J, et al. Our future in the anthropocene biosphere. *Ambio.* 2021;50(4):834–869. https://www.ncbi.nlm.nih.gov/pmc/articles/PMC7955950/pdf/13280_2021_Article_1544.pdf.
46. Wilson EO. *Half-Earth: Our Planet's Fight for Life.* New York, NY: Liveright; 2006.
47. Wikipedia. *Half-Earth.* https://en.wikipedia.org/wiki/Half-Earth.

48. Beyer RM, Manica A, Mora C. Shifts in global bat diversity suggest a possible role of climate change in the emergence of SARS-CoV-1 and SARS-CoV-2. *Sci Total Environ*. 2021;767, 145413. https://www.sciencedirect.com/science/article/pii/S0048969721004812?via%3Dihub.
49. Moore S. Climate change and COVID-19. News medical life. *Science*. 2021. https://www.news-medical.net/health/Climate-Change-and-COVID-19.aspx.
50. Bresson D. Climate change could have played a role in the Covid-19 outbreak. *Forbes*. 2021. Available from: https://www.forbes.com/sites/davidbressan/2021/02/08/climate-change-could-have-played-a-role-in-the-covid-19-outbreak/?sh=3c4240b611ef.
51. Zang SM, Benjenk I, Breakey S, et al. The intersection of climate change with the era of COVID-19. *Public Health Nurs*. 2021;38(2):321–335. https://www.ncbi.nlm.nih.gov/pmc/articles/PMC8014081/pdf/PHN-9999-0.pdf.
52. Muller JE, Nathan DG. COVID-19, nuclear war, and global warming: lessons for our vulnerable world. *Lancet*. 2020;395(10242):1967–1968. https://www.ncbi.nlm.nih.gov/pmc/articles/PMC7292599/pdf/main.pdf.
53. Schwartz SA. Climate change, Covid-19, preparedness, and consciousness. *Explore*. 2020;16(3):141–144. https://www.ncbi.nlm.nih.gov/pmc/articles/PMC7102555/pdf/main.pdf.
54. Barouki R, Kogevinas M, Audouze K, et al. The COVID-19 pandemic and global environmental change: emerging research needs. *Environ Int*. 2021;146, 106272. https://www.ncbi.nlm.nih.gov/pmc/articles/PMC7674147/pdf/main.pdf.
55. European Health Emergency Preparedness and Response Authority (HERA). www.heraresearcheu.eu.
56. European Health Emergency Preparedness and Response Authority (HERA). https://ec.europa.eu/info/law/better-regulation/have-your-say/initiatives/12870-European-Health-Emergency-Preparedness-and-Response-Authority-HERA-_en.
57. European Health Emergency Preparedness and Response Authority (HERA). https://de.wikipedia.org/wiki/European_Health_Emergency_Response_Authority.
58. *HERA Work Packages*. https://www.heraresearcheu.eu/workpackages.
59. WHO. *Declaration of Alma-Ata*; 1978. https://www.who.int/publications/almaata_declaration_en.pdf.
60. EU Sustainable Development Goals (SDGs). https://www.euro.who.int/en/health-topics/health-policy/sustainable-development-goals.
61. European Commission. *Sustainable Development Goals (SDGs)*. https://ec.europa.eu/international-partnerships/sustainable-development-goals_en.
62. Editorial The Lancet. Climate and COVID-19: converging crises. *Lancet*. 2021;397(10269):71. December 2020 https://www.thelancet.com/journals/lancet/article/PIIS0140-6736(20)32579-4/fulltext.
63. *Impacts of COVID-19 disproportionately affect poor and vulnerable: UN chief*. UN News; 2020. Available from: https://news.un.org/en/story/2020/06/1067502.
64. Malthus TR. *An Essay on the Principle of Population*. Oxford, UK: Oxford University Press; 2008.
65. Wikipedia. *Economics*. https://en.wikipedia.org/wiki/Economics.
66. Wikipedia. *Thomas Robert Malthus*. https://en.wikipedia.org/wiki/Thomas_Robert_Malthus.
67. Ciesla WM, Donaubauer E. *Decline and Dieback of Trees and Forests: A Global Overview (FAO Forestry Paper 120)*. UN Food and Agriculture Organisation (FAO); 1994. http://www.fao.org/3/ap429e/ap429e00.pdf.
68. Wikipedia. Club of Rome. https://en.wikipedia.org/wiki/Club_of_Rome.
69. Meadows DH, Meadows DL, Randers J, Behrens III WM. *The Limits to Growth*. New York, NY: Universe Books; 1972. http://www.donellameadows.org/wp-content/userfiles/Limits-to-Growth-digital-scan-version.pdf.

70. King A, Schneider B. *The First Global Revolution. A Report by the Council of the Club of Rome.* New York, NY: Pantheon Books; 1991. https://img1.wsimg.com/blobby/go/a437931e-3d50-4da5-a130-ac99e488e617/downloads/The%20First%20Global%20Revolution_%20A%20Report%20by%20the%20C.pdf?ver=1624377297299.
71. von Weizsäcker EU, Wijkman A. *Come on! Capitalism, Short-Termism, Population and the Destruction of the Planet.* New York, NY: Springer; 2017.
72. *The Club of Rome Supports Global Student Climate Protests.* Club of Rome; 2019. https://www.clubofrome.org/impact-hubs/climate-emergency/the-club-of-rome-supports-global-student-climate-protests/.
73. *Statement in Support of global Student Climate Protests.* Club of Rome; 2019. https://clubofrome.org/wp-content/uploads/2020/03/Club-of-Rome-statement-on-student-protests-18.03.pdf.
74. *Club of Rome Summit 2019: Reclaiming & Reframing Economics.* https://www.youtube.com/watch?v=UDihQdQu43k.
75. Cole HSD. *Models of Doom: A Critique of the Limits to Growth.* New York, NY: Universe Publishing, Simon & Schuster; 1973.
76. Alpert JS. If you are not a liberal when you are young, you have no heart, and if you are not a conservative when old, you have no brain. *Am J Med.* 2016;129(7):647–648. https://www.amjmed.com/article/S0002-9343(16)30193-0/pdf.

CHAPTER

16

China and Russia are giants on feet of clay

The expression of a giant on feet of clay derives from the interpretation of a dream of Nebuchadnezzar, King of Babylon, by the prophet Daniel as recounted in the Bible. In his dream, Nebuchadnezzar had seen a giant statue, awesome in appearance, but on closer inspection with feet partly of iron, and partly of baked clay. Daniel prophesied that comparable to the feet of clay, the kingdom of Babylon would be both strong and fragile. To this day, the expression of a giant on feet of clay is used to describe institutions that appear strong and invincible, but have weaknesses that are invisible at first glance. Both Russia and China are great states with great past and great aspirations for a leadership role in today's world. Both tried to establish themselves as heavyweights in the fight against the COVID-19 pandemic. As we have shown in the previous chapters, the self-perception of these two states does not match the reality.

In Russia, the centuries-long rule of the tsar was replaced at the end of the First World War by the rule of the communists, who murdered the tsarist family (including their servants, who were not noble) and established the "dictatorship of the proletariat" in the political form of a soviet republic (Union of Soviet Socialist Republics, USSR). The USSR emerged from the Second World War as a superpower, with the counterbalance of the USA, the western European states, and the rest of the free world. When the Soviet Union imploded, many former parts became independent states, while Russia succeeded it and inherited its arsenal of nuclear weapons. But Russia has not yet recovered from the post-Soviet legacy of the planned economy. Already in the years before the dissolution of the Soviet Union, the economy was in a state of decline. Official statistics masked industrial inefficiencies. After the implosion of the USSR in 1991, Russia implemented reforms to transform the economy from central planning and control toward a market-driven approach, including the establishment of privately owned industrial and commercial ventures and privatizing state-owned enterprises. But the privatization process was slow. Many heavy industries remained under state control. Foreign direct investment was encouraged in principle but was constrained by unfavorable conditions, state interventions in industry, corruption, and weakness of the legal framework. Furthermore, the full extent of serious long-term environmental degradation during the Soviet period became apparent in the 1990s. Under the presidentship of Vladimir Putin, Russia became again an authoritarian state. To a relevant extent, the past decades have been lost years from an economic point of view. Russia is the largest country in the world by land surface, but its economy continues to depend heavily on the export of oil, gas, and other raw materials. Russia is one of the largest arms exporters in the world and still has large and strong-armed forces, but its economic base is comparatively small. It is often described as an Upper Volta with nuclear weapons, a third world state with atomic bombs. The USSR had become a Potëmkin state like the Potëmkin villages of the eighteenth century, built by the Russian Prince Grigory Potëmkin, who constructed fake facades to mimic real villages and well-fed people for the tsar and his entourage to see. Russia's current ruler dreams of the mighty tsarist empire of the past. But its gross domestic product (GDP) ranks behind that of France, Italy, or Canada. It has a population of over 140 million residents and has well-functioning armed forces that it uses to interfere in conflicts around the world to continue to have a say in world politics. There is no longer any ideology that governs the Russian government. Officially, it is a democracy, but in fact, it is a dictatorship of those who were washed by cunning and history into leadership and government. Elections do not serve to elect (and vote out) politicians. They are a facade that hardly anyone in the population believes anymore.[1-5]

In early history, China had a leading position in science and technology. It was one of the nuclei of world civilization that developed writing. Papermaking, printing, the compass, and gunpowder were first developed in China and partially reached Europe centuries after their development. Chinese scientific leadership started to decline in the 14th century. After having been defeated repeatedly by Japan and European nations in the 19th century, Chinese reformers began promoting modern science and technology. When the Communist Party gained power in 1949, science and research were organized based on the model of the Soviet Union. Key characteristics were their bureaucratic organization, central plans, separation of research from production, specialized research institutes, concentration on practical applications, and restrictions on information flow. Many studied in the Soviet Union which also transferred technology.[6-8]

After its economic reforms in 1978 and its entry into the World Trade Organization (WTO) in 2001, China transitioned from a planned economy to a mixed economy with an increasingly open-market environment. China's economy became the second-largest country by nominal gross domestic product (GDP) in 2010 (GDP is the market value of all goods and services produced in a specific time) and grew to the largest in the world by purchasing power parity (PPP) in 2014 (PPP compares different countries' currencies through a "basket of goods" approach). China is today the world's fastest-growing major economy. It is run by the Chinese Communist Party. In theory, the party's ideology is still the teachings of Marx, Engels, Lenin, and Stalin, although this official ideology stands in contrast to today's market orientation of the economy. Furthermore, China is sliding back into a nationalist position and is militarily becoming more and more aggressive toward its neighbors and the world. The Chinese constitution states that the People's Republic of China "is a socialist state governed by a people's democratic dictatorship that is led by the working class and based on an alliance of workers and peasants," and that the state institutions "shall practice the principle of democratic centralism." The Constitution of the People's Republic of China states that the "fundamental rights" of citizens include freedom of speech, freedom of the press, the right to a fair trial, freedom of religion, universal suffrage, and property rights. However, in practice, these provisions do not afford significant protection against criminal prosecution by the state. Although some criticisms of government policies and the ruling Communist Party are tolerated, censorship of political speech and information, most notably on the Internet, are routinely used to prevent collective action. By 2020, China plans to give all its citizens a personal "Social Credit" score based on how they behave. The Social Credit System, now being piloted in several Chinese cities, is considered a form of mass surveillance that uses big data analysis technology. Its current political, ideological and economic system has been termed by its leaders as a "consultative democracy" "people's democratic dictatorship," "socialism with Chinese characteristics" (which is Marxism adapted to Chinese circumstances), and the "socialist market economy," respectively. There are no freely elected national leaders, political opposition is suppressed, religious activity is controlled by the state, dissent is not permitted and civil rights are curtailed.[8–11]

Both China and Russia have only a very limited tradition of free scientific publishing. The interplay between the pharmaceutical companies, which are in tough competition with one another, the regulatory authorities who set clear rules for the clinical trials that lead to the registration of drugs and vaccines, and the public debates that lead to regular modifications of the state framework are necessary for drug development.[12] All these elements are absent in both countries. The interplay in the power triangle of the pharmaceutical industry, regulatory authorities, and clinical medicine, over which the rules enacted by the lawmakers stand as a framework for action, has led to a structure in the USA for more than half a century. Drug development is not just a technical achievement. It is part of society, even if it is currently rather underestimated in the West because certain academic circles enjoy regularly attacking the pharmaceutical industry for its alleged greed for money.[13–15]

Scientific advancement and innovation are inextricably linked with freedom of expression and the opportunity to develop professionally without the state controlling everything.

The Chinese economy is still catching up with the developed western industrial nations. At the formal level, government planning decisions are much easier in China. If the country needs a bigger airport, new metro lines, more warships, or more spaceships, they will be designed and built. But catching up with superior scientific standards and independent scientific and technical innovation are two different things. For the development of COVID-19 vaccines, both Russia and China relied on traditional methods that in the COVID-19 pandemic proved to be outdated and too poorly effective. Furthermore, there was no sufficient transparency of clinical data.

Chinese society has always had to deal with restricted spaces for advocacy. The purpose of censorship in China is not to silence all comments made about the state or any particular issues, but to prevent and reduce the probability of collective actions. Allowing social media to flourish also has allowed negative and positive comments about the state and its leaders to exist. Civil society advocacy is possible to some degree as long as it does not lead to collective action.[11]

The Chinese penal system includes labor prison factories, detention centers, and re-education camps. US Research estimated that there were over a 1000 slave labor prisons and camps, known collectively as the Laogai.[8]

Science and technology in China have developed rapidly over the past 20 years. The Chinese government perceives science and technology as a fundamental part of socio-economic development. The country has made rapid advances in education, infrastructure, high-tech manufacturing, academic publishing, patents, and commercial applications. It is now increasingly targeting its own innovation and aims to reform remaining weaknesses. Prioritized industries and firms are protected and guided. There are systematic efforts to replace foreign technology and intellectual properties with own, homemade technology. Foreign companies were given many incentives for technology transfer and for moving research and development (R&D) to China. At the same time, the technological abilities of Chinese companies are supported in various ways. Nationalism has become the main ideological justifications and societal glue for the regime, although Marxism continues to be maintained as the official doctrine. Some science and technology mega-projects might be seen as trophy projects or propaganda purposes with Chinese state-controlled media being filled with reports of Chinese achievements, such as the advances of Chinese space exploration.[7–9] Nevertheless, China's space technology is still a follow-up to the Soviet Union's technology and is state-organized, while the US is now revolutionizing space travel with private companies. Transports commissioned by the USA to the international space station will continue to be coordinated by NASA, but the transports themselves are now being carried out by companies such as SpaceX and others.[16-19]

Science and R&D are at odds with the culture of a police state where observations made by caring physicians are monitored and censored by the police.[20] The official story of the efficacy of the vaccines developed and produced by Sinovac and Sinopharm is now at odds with clinical experiences made worldwide outside the sphere of influence of the Chinese surveillance apparatus. When the director of the China Centers for Disease Control and Prevention (CDC) admitted that the Chinese vaccines' efficacy is relatively low and that the government is considering mixing them to give them a boost, he will certainly have received permission from his political controllers to say so in public.[21,22]

Both Russia and China had hoped to win sympathy worldwide, specifically in developing countries that initially had no access to Western vaccines, by offering their vaccines, developed and produced at home. Now, in the second year of the pandemic, and considering the high efficacy of the vaccines approved by FDA, EMA, MHRA, and PMDA, the world is beginning to see the divergence of appearances and realities when looking at the claims of the leadership of these two countries. Neither Russian nor Chinese companies have been sufficiently transparent about their trial data.[23]

References

1. Paul Bushkovitch. *A Concise History of Russia (Cambridge Concise Histories).* Cambridge, UK: Cambridge University Press; 2012.
2. Captivating History. *History of Russia: A Captivating Guide to Russian History, Ivan the Terrible, The Russian Revolution and Cambridge Five.* Redhill, UK: C H Publications; 2019.
3. Britannica. *Economy of Russia.* https://www.britannica.com/place/Russia/Economy.
4. Wikipedia. *Comparison Between U.S. States and Sovereign States by GDP.* https://en.wikipedia.org/wiki/Comparison_between_U.S._states_and_sovereign_states_by_GDP.
5. Russia does not want to be 'Upper Volta with nuclear weapons', says Biden. Frontier India n.d. https://frontierindia.com/russia-does-not-want-to-be-upper-volta-with-nuclear-weapons-says-biden/.
6. Wood M. *Story of China.* New York, NY: Simon & Schuster; 2020.
7. Wikipedia. *Science and Technology in China.* https://en.wikipedia.org/wiki/Science_and_technology_in_China.
8. Wikipedia. *China.* https://en.wikipedia.org/wiki/China.
9. Weber IM. *How China Escaped Shock Therapy: The Market Reform Debate.* London, UK: Routledge; 2021.
10. Strittmatter K. *We Have Been Harmonized. Life in China's Surveillance State.* New York, NY: Harper Collins; 2020.
11. Wikipedia. *Politics of China.* https://en.wikipedia.org/wiki/Politics_of_China.
12. Hilts PJ. *Protecting America's Health: The FDA, Business, and One Hundred Years of Regulation.* New York, NY: Alfred A Knopf Publishers; 2003.
13. Angell M. *The Truth About the Drug Companies: How They Deceive Us and What to Do About It.* New York, USA: Random House Publishers; 2004.

14. Goldacre B. *Bad Pharma: How Drug Companies Mislead Doctors and Harm Patients.* London, UK: Harper Collins Publishers; 2012.
15. Gøtzsche P. *Deadly Medicines and Organised Crime: How Big Pharma Has Corrupted Healthcare.* London, UK: Routledge; 2013.
16. SpaceX. https://www.spacex.com/.
17. Wikipedia. *SpaceX.* https://en.wikipedia.org/wiki/SpaceX.
18. Northrop Grumman. https://www.northropgrumman.com/.
19. Wikipedia. *Northrop Grumman.* https://en.wikipedia.org/wiki/Northrop_Grumman.
20. Wikipdedia. *Li Wenliang.* https://en.wikipedia.org/wiki/Li_Wenliang.
21. McDonald J, Wu H. *Top Chinese Official Admits Vaccines Have Low Effectiveness.* AP News; April 11, 2021. https://apnews.com/article/china-gao-fu-vaccines-offer-low-protection-coronavirus-675bcb6b5710c7329823148ffbff6ef9.
22. McGregor G. *1.4 Billion Doses Later, China Is Realizing It May Need mRNA COVID Vaccines.* Fortune; July 17, 2021. https://fortune.com/2021/07/16/china-mrna-vaccine-pfizer-biontech-fosun-doses/.
23. West J. *China's Innovation Dilemma. Tech Advances Have Been Impressive, but Constraints on Business Productivity and Imagination Are a Major Obstacle*; May 10, 2021. https://www.lowyinstitute.org/the-interpreter/china-s-innovation-dilemma.

CHAPTER

17

Conclusions and outlook

The COVID-19 pandemic had far-reaching economic consequences beyond the spread of the disease itself and efforts to contain it when the SARS-CoV-2 virus spread around the globe. It caused the second largest global recession in history, after the great depression that started in 1929 in the USA, became worldwide and lasted until the late 1930s. The recession triggered by the COVID-19 pandemic was and is unique in many ways. The contraction in 2020 was very sudden and deep compared to previous global crises. The recession also stands out for its differential impacts across sectors and countries. During the pandemic, more than a third of the global population was in lockdown. It affected almost all parts of the economy and with that the entire society, including manufacturing, agriculture, arts, publishing, retail, hospitality business, tourism, aviation, cruise lines, railways, and more. Millions of jobs were lost.[1-3]

The COVID-19 pandemic is often compared to a war. Although it has not been World War III, this comparison should not be completely dismissed. With war, we instantaneously associate horror, destruction, and the senseless annihilation of human life on a large scale. But wars have also other sides. They not only have winners and defeated. They also destroy intellectual constructs that were partly responsible for their emergence. The two world wars brought down Royal families in Europe, who had competed over the previous centuries about who was richer, more sophisticated, bigger, and more beautiful. These centuries created great treasures in painting, music, science, poetry, literature, and more, but at a high price for those who had no voice. In the end, the European Countries had almost the whole world at their disposal, and this competition culminated in the first two world wars. But these wars also made it easier for women to enter public life because the men were initially in military service or at the front, and later many did not return. It was not just the physical shortage of manpower. The mindsets that had suppressed the self-determined presence of women in the workplace and the public, in general, had been crushed along with the physical destructions. The two world wars resulted in many democracies around the world, albeit, again, at a terrible price. They resulted in worldwide decolonization in the decades thereafter. The world wars also allowed experiments at the state level that decades later can help us open our eyes. The unlimited rule of socialism without controls has shown with the Soviet Union that behind the beautiful words and promises there is the naked lust for power that has brought about more serious crimes than ordinary people who had fought for their ideals could have ever dreamed of. Eventually, the Soviet Union imploded and was inherited by a government of Russia that dreams of the past greatness of tzarism. It uses its considerable military strength to be taken seriously on the international stage, ruthlessly, in an unimaginable series of war crimes. With its National Socialism, Germany shares with the Soviet Union the sad honor of having committed among the worst crimes in human history. Today we face a new emerging superpower that draws its inner strength from the teachings of Marx, Lenin, and Stalin, now continued by the new prophets Mao Zedong and Xi Jinping. But our world today is no longer the world after the destruction of the two world wars. Not only have new technologies emerged, but also new ideas have taken hold. But resting on the laurels of past successes is always a bad approach.

Wars mobilize both the most brutal and primitive instincts of man, as well as the noblest, especially when it comes to cohesion and help for others, but sometimes also in grace toward the defeated. They allow a collective effort in which technical innovations can prevail, which under normal circumstances would take decades or longer to prevail. World War II brought antibiotics, radar, and nuclear weapons. The decades thereafter brought artificial intelligence and the internet. The market-oriented, western world is colorful, crazy, and full of contradictions. It allows individuals to speak up without being immediately imprisoned or worse for insubordination. The COVID-19 pandemic has brought us to a point where we will have to tackle the next roadblocks to advance humanity even further technically and culturally. And we will have to continue defending ourselves against ideas and ideologies from the moth box of the past, which from inside and outside are continuously trying to get their cold hands on the throat of western society and freedom. We all need to continually learn to distinguish between what people and institutions *say* and what they *mean*.

Humanity was surprised by the pandemic. To a certain degree, countries were prepared for large accidents or major natural disasters, with huge differences between countries. Some countries had reserves and emergency supplies while others have less or none at all. And none was prepared for a global viral pandemic. In developed countries, it had been assumed that infectious diseases would threaten people predominantly in developing countries. Infectious pandemics were regarded as something from the past. The developed world has been successful in fighting infectious diseases. Primarily not through the development of antibiotics, but elementary sanitary measures such as the provision of clean drinking water, effective removal of harmful wastewater, higher hygiene standards, better nutrition, better housing, and more. Malaria in Europe has been largely suppressed by draining swamps. Antibiotics and later other anti-infectives then helped those who were infected nevertheless. After World War II, drug development exploded and expanded. The chemical industry became pharmaceutical, then biopharmaceutical, and finally life sciences industry.

The developed countries have not only become inert, which is to be expected from long periods of success. Additional elements have slipped in. New drugs or vaccines will not save lives because they were invented. To help patients, they must first be developed, then approved, then produced, stored, and distributed to pharmacies. Only then can they be administered. Not only authorities are involved in the approval of drugs and active ingredients. They get their advice from independent scientists and their associations. Behind all of these factors is the general public, which through elections has an elementary, albeit very indirect, influence on health policy decisions. The general public, in turn, is influenced by the world of communication, which has taken on a life of its own with growing literacy and expanding technological means, including print media, radio, television, now the Internet, and finally social media.

Values, ideas, and illusions that are more in line with the world of the 19th century still play a key role in this interpersonal world of communication. The caring medical doctor and the caring nurse are characterized as the white knights, and as a counterbalance to the pharmaceutical industry, which allegedly is greedy and only wants to increase its profits.[4-6] These values are still extremely strong in Europe, while the USA has found a more pragmatic and realistic business relationship between bedside care, its administration, and the research and development that continuously advances medical treatment further.

It was these traditional values and the self-image of academic clinicians that shaped the response to the COVID-19 pandemic. They are particularly strong in Europe. The first textbooks on the COVID-19 pandemic were written by terribly important professors (TIPs) in medicine and related disciplines. Once you suffer from an illness, you must indeed trust the treating physician. But the most important task in modern society and public health is to prevent diseases as much as possible. This has worked pretty well in the past for infectious diseases. And it was the challenge mankind faced and still is facing with the COVID-19 pandemic. It hit a highly developed global civilization that is increasingly embracing high technology. Without modern transport systems, international business, tourism, trade, and many other factors, the pandemic would not have spread so fast. A few 100 years ago, the same virus would probably, if it had spread at all, not even been recognized as a challenge. It was science, applied research, and its translation into industrial production of high-tech vaccines that now allow effective prevention of COVID-19. The genome of the virus was deciphered by Chinese scientists within a few days and uploaded to an internationally accessible database. Without genetic engineering, we could have done little with knowing of the spikes on the surface of the virus. Without messenger RNA, we would not be able to make the human body produce the spike protein itself, preparing the immune system for future encounters with the invading viruses and destroy them before they can penetrate too many cells and cause damage.

At the political level, most governments initially had difficulties recognizing the seriousness of the situation. And even if they recognized it, they had only very limited passive measures at their disposal: tracking up infections, isolating the infected, and quarantining those who had come into contact with infected people. These classic measures have worked relatively well in many countries, especially island states like Australia or New Zealand, where geography made this easier. A special case was China, where the pandemic had broken out. The Chinese leadership imposed draconian travel restrictions and lockdowns on a scale that would have been unimaginable in free countries. At the technical level, this approach was successful as far as we know. And as the free press does no longer exist in China, and reporters from free countries are no longer welcome, we will never learn details that the Chinese leadership does not want to learn about. Furthermore, even those countries that initially fared rather well with preventing the pandemic from spreading later lagged with the vaccination of their citizens. This is of increasing relevance with the spread of virus variants that are significantly more contagious than the original virus.

The US reaction was intriguingly twofold. President Donald Trump did not realize and acknowledge the severity of the pandemic until the very end of his presidency. In addition, political forces of all kinds denied the need for protective measures and allowed the pandemic to spread further than would have been necessary. But the other side was that the US is the leading country in modern biotechnology and that the leaders and institutions involved in biotechnology managed to move the US President to take a reasonable step. Operation Warp Speed (OWS) freed up relevant funds for the development and early ordering of vaccines, drugs, and diagnostics against COVID-19 at record speed. When Trump was voted out, the successor government had enough vaccines at hand. Although initially intended for the United States only, OWS made vaccines, drugs, and diagnostics available worldwide, albeit only after meeting the United States' urgent needs. On a technical and financial level, also a union like the EU or a country like Germany would have had the same option. And what did Germany do? It waited for the EU to order vaccines. And the EU? It took its time.

There were also several other governments whose response to the pandemic can only be described in terms of children's circus categories. To this day, the President of Brazil does not believe that the pandemic should be taken seriously. Russia has continuously glossed over the number of infections in its own country. I will not list all the countries of the world here. Suffice it to say that there was a patchwork of government reactions ranging from reasonable to openly idiotic.

The European Union (EU) fundamentally neglected the duty of any government to look after its citizens first. There was much reliance on medical experts and academic researchers combined with a fundamentally flawed view of the relationship between public health, bedside treatment, and the development of drugs and vaccines. There was also a parsimonious attitude at the level of the governments of the EU countries, who hoped to save money when ordering vaccines by bargaining hard and ordering as cheap as possible. The other side of the coin was that the EU supported the COVID-19 Vaccines Global Access (COVAX) initiative to give the developing world access to vaccines. This was nice and politically correct. But it does not establish an international leadership position. The EU and the European individual national governments should have immediately promoted the development of new vaccines on a level comparable to that of the US and should have ordered them in advance. They should have thought of their citizens first. No government can survive for long if it puts the rest of the world above its people. With COVAX, the EU praised international solidarity after global vaccine development had already gotten underway. But you cannot stand next to a train, wait until it has passed, and then claim that you are the leader of that train. Today, the USA is by far the largest donor to COVAX.

Educated people and scientists who were and are aware of the limitations of individual governments often place great trust in the World Health Organization (WHO), which is an international authority in health care on paper and in theory. The WHO is responsible for the international epidemiological monitoring of diseases and pandemics. But it has neither the financial resources, nor the mandate, nor the industrial fundament to actively coordinate the fight against pandemics. And it has a flawed self-image, visible from the statements made by its Director-General and from the many politically correct representations on the numerous WHO websites. The basis of science and knowledge is not the opinions expressed by a majority of citizens, countries, governments, or institutions. And the basis for industrial production is far more than politically correct declarations.

Why was there no global initiative to facilitate the development of vaccines, drugs, and diagnostics at the beginning of the pandemic, jointly by e.g., the USA, the EU, Japan, China, and other countries? It sounds nicer and more politically correct to invoke the common interests of humanity. But there is also competition and conflicts of interest. Not only secret conflicts of interest between institutions in the free world, such as the desire to secure jobs in science based on pointless "pediatric" studies,[7-11] but also those between states, and more. Nationalism and national pride continue to be powerful drivers in international affairs, and they have their ugly sides, too. Perhaps the EU is the only region in the world that is fed up with nationalism to a relevant degree. The devastating experiences of the Second World War shape the collective memory of Europeans to this day, albeit except for of a few small states that we're able to avoid the Armageddon of destruction, such as Switzerland and Sweden.

The scientific discussion of the COVID-19 pandemic and publications about it have given us essential keys to understanding the pandemic. Eventually, this will hopefully help us also to advance further with the prevention and treatment of many further diseases. But science and its institutions are also in a crisis. A blur has emerged between the contents and the aims of science and the material interest of scientists, as well as the self-interests of scientific institutions that *say* they want to advance science, but *mean* something else. Increasingly, the purpose of scientific publishing is no longer the desire to advance knowledge and understanding, but to advance personal careers for which scientific (and pseudo-scientific) publications have become the currency that is needed to advance. In many specialized areas, progress continues. The number of journals is exploding, and among all the published papers are more and more banal ones. All this is reinforced through the internet.

Two areas of science are specifically addressed in this book, in which more and more studies are carried out and published that have less to do with gaining knowledge and more with enabling more and more scientists to take up academic positions. These are pointless and harmful clinical studies with minors, and studies on global warming. It is not being said here that global warming does not exist. But the demand to promote more and more scientific research in this area will result in more and more banal observations being published without anything being achieved. There are certainly many more areas. But this is another story.

This situation puts people who try to get their picture in a difficult position. Blind trust in one's government was really out of place, with the possible exception of the citizens of Israel. Israel obtained effective vaccines as early as possible and vaccinated the population to a large extent very quickly. Israel also had to go through a temporary lockdown, and Israel is also struggling with the Delta variant. But Israel has quickly started giving people at risk a third dose of the vaccine and has the pandemic under control better than most other countries.[12-16]

The pandemic is a challenge that goes well beyond just medical and health issues. The complexity of this challenge is unique. As with most crises, it will take time for the public and scientific debate to reach reasonable conclusions.

17. Conclusions and outlook

The artificial separation of humanity into two populations, those above and those below the 18th birthday, in medicine and drug approval, is a large area in which traditional structures and thought structures interfere with medical prevention and treatment. Administratively and legally, we have to differentiate between adults and minors. A minor will be admitted to a pediatric hospital. But he should receive the same diagnosis and the same treatment for the same diseases before and after the 18th birthday. But our society blurs physiological and administrative/legal criteria. Questionable pediatric studies are formally based on a blur between humans as minors for administrative/legal purposes and the fact that they mature long before the 18th birthday. Adolescents could have been vaccinated against COVID-19 much earlier. The 18th birthday is a legal and legal limit, not a physical one. Medicine deals with people, not their legal status. The deeper roots of the artificial separation of humanity into two populations are hidden conflicts of interest. At the beginning of the pandemic, very few minors got infected. Now, younger patients are increasingly infected by the more infectious virus variants. In the US, there are several reports of children that have died of COVID-19.[17–20] With that the question arises why minors are not vaccinated now, including young ones, except for newborns and babies, where additional caution is necessary.[7–11] For vaccinating minors under the age of 12 or under 5 years, it is nonsense to demand separate efficacy studies. The vaccines work before and after the 12th or 5th birthdays. But worldwide, the regulatory authorities demand separate efficacy studies. There is a conglomerate of scientists, scientific institutions, employees of regulatory authorities, and employees of international organizations who all benefit from separate "pediatric" studies and have established their careers on them, even if the alleged "children" are already bodily mature. Blind trust in institutions that were successful some decades earlier is a challenge of the 21st century. In the past, it often took centuries before critical questions were asked. Mankind will face many more challenges in the next decades, centuries, and millennia. Pediatrics is historically a very young clinical discipline. When it emerged, there were hardly any effective modern drugs. Diseases such as cancer, joint inflammation in minors, septicemia, or cystic fibrosis were incurable or left young people and their parents with years of misery. The 18th birthday was largely irrelevant. Things have changed. Today everyone carries an ID; babies have already their passports; and a birthday, not physical maturity, determines whether you can be vaccinated against COVID-19. Hardly anyone dares to question that something could be wrong with the administrative age limits for medical interventions.[7–11] Let us hope the pandemic and the reflections while it is gradually overcome will help quicken the pace of processing.

All this leaves the task of finding a compass in the COVID-19 pandemic for the skeptical individual and critical scientists, journalists, patient representatives, governments, and others. And maybe those who did most of the work to ensure that we see a light today at the end of the COVID-19 pandemic tunnel: the companies that successfully developed the vaccines against the COVID-19 virus.[21–23]

In the free world, nobody is thrown into prison for addressing flaws and errors committed by governments. Go to Russia or China and express such seditious thoughts. You will go to jail, will have a tragic accident, or simply disappear, sometimes with a preceding warning, sometimes without.

1 Conclusions

The COVID-19 pandemic is a challenge to our trust in science, authorities, and the structures we live in. All relevant organizations and structures in healthcare will have to review themselves and be reviewed by the public. Either the professional organizations, representations, and structures will address this themselves, or pressure from the public will have to help.[7–11]

Health is an area where high technology meets every single individual. The value of vaccinations tends to be forgotten when infectious diseases seem to have been overcome. Many people can no longer imagine the potential horrors of infectious diseases. The COVID-19 pandemic should encourage us to put the relationship between health and high technology on a new basis. We learn too many useless things in school and too little about the essentials of modern life. It would be desirable for health and health challenges to become part of compulsory education. This should include teaching how vaccines work, major diseases such as diabetes, epilepsy, Parkinson's disease, Alzheimer's, and several more; how drugs are developed and authorized; and several additional issues. The pharmaceutical industry and the representative bodies of the healthcare professions could and should contribute. The industry should be encouraged to use the opportunity to teach about applied science in schools and universities. The pharmaceutical and life science industry should be encouraged to go out of the defensive and get rid of being portrayed as the villain.[4–6] This should not and must not result in hidden advertising. But modern drugs and vaccines have fundamentally changed our world and the composition of mankind. People with diabetes, cystic fibrosis, and a myriad of further, previously lethal diseases now live a long life. Parents do no longer hide their handicapped children. Healthcare and modern drugs concern everyone, and the earlier young people learn about applied research in healthcare, the fewer confidence tricksters and other disbelievers in science and innovation have later a chance that the nonsense they spread will also be believed.

There are more ways for the life science industry to get out of its defensive position. Recently, Stéphane Bancel, chief executive officer (CEO) of Moderna, the manufacturer of one of the FDA-approved mRNA vaccines, shared in an interview his vision of the future of the SARS-CoV-2 virus. In his view, the coronavirus pandemic could be over by the end of 2022 as increased vaccine production is ensuring global supplies. He compared the future SARS-CoV-2 position with the flu: people can either get vaccinated and have a good time, or they refuse vaccination and risk getting sick and ending up in the hospital, or worse.[21-23] Maybe Bancel is slightly too optimistic, but it is just good not to hear for once just another professional complainer.

The roots of the demonization of the life science industry should be addressed more offensively. Researchers and others that demand pointless and harmful "pediatric" studies in adolescents and have so far prevented minors to be vaccinated against COVID-19, never tire of condemning the industry as being greedy. But they are greedy, having based their careers on conducting pointless and harmful "pediatric" studies. The bodies representing the medical professions, including the American Medical Association (AMA), the American Academy of Pediatrics (AAP), internal medicine, and many other sub-specialties should distance themselves from age limits that define "children" from adults for medical interventions. Institutional Review Boards (IRBs)/ethics committees (ECs) should suspend questionable "pediatric" studies and should reject submitted new ones.[7-11] The International Committee of Medical Journal Editors (ICMJE) should distance itself from questionable "pediatric" studies and should adapt its recommendations for the conduct, reporting, editing, and publication of scholarly work in medical journals accordingly.[24] It could do so for all "pediatric" studies, not just for those related to COVID-19.

The free world will need to get rid of many illusions and will have to make its dealings with other countries and institutions more rational. Donald Trump's criticism of the WHO during the first few months of the pandemic was factually justified to a relevant degree. The way in which he withdrew the US from the WHO, however, was his typical behavior of the elephant that got lost in the China shop. All countries must deal with the WHO and use its epidemiological services, its international classification of diseases (ICD),[25-27] and more. However, it is wrong to take the WHO's recommendations, pronouncements, and requests at face value. We will have to learn to distinguish between our desires and illusions, the harsh reality, and the self-image of the WHO bureaucrats and the governments that delegate them. Being diplomatic, but also knowing precisely what to think about the respective counterpart is nothing new in diplomacy. However, modern times and modern communication, specifically internet-based communication of international organizations and they are being mirrored in social media, has given them a higher power than deserved. In relation to the WHO and many other international organizations and institutions, diplomacy will always be necessary. And it will be necessary to say goodbye to the illusion that the WHO has more reliability and scientific authority in health issues because it represents the majority of the world's governments and public health institutions.

References

1. Barrett P, Das S, Magistretti G, et al. *After-Effects of the COVID-19 Pandemic: Prospects for Medium-Term Economic Damage.* International Monetary Fund Working Papers; July 30, 2021. https://www.imf.org/en/Publications/WP/Issues/2021/07/30/After-Effects-of-the-COVID-19-Pandemic-Prospects-for-Medium-Term-Economic-Damage-462898.
2. Wikipedia. *Economic Impact of the COVID-19 Pandemic.* https://en.wikipedia.org/wiki/Economic_impact_of_the_COVID-19_pandemic.
3. Wikipedia. *COVID-19 Recession.* https://en.wikipedia.org/wiki/COVID-19_recession.
4. Angell M. *The Truth about the Drug Companies: How they Deceive us and What to Do about it.* New York, USA: Random House Publishers; 2004.
5. Goldacre B. *Bad Pharma: How Drug Companies Mislead Doctors and Harm Patients.* London, UK: Harper Collins Publishers; 2012.
6. Gøtzsche P. *Deadly Medicines and Organised Crime: How Big Pharma Has Corrupted Healthcare.* London, UK: Routledge; 2013.
7. Rose K. *Blind Trust- How Parents With a Sick Child Can Escape the Labyrinth of Lies, Hypocrisy and False Promises of Researchers and Regulatory Authorities.* London, UK: Hammersmith; 2021. https://www.chapters.indigo.ca/en-ca/books/blind-trust/9781781612026-item.html. https://www.amazon.com/-/de/dp/1781612021/ref=sr_1_2?dchild=1&qid=1620137631&refinements=p_27%3AKlaus+Rose&s=books&sr=1-2.
8. Rose K, Grant-Kels JM, Etienne E, Tanjinatus E, Striano P, Neubauer D. COVID-19 and treatment and immunization of children—the time to redefine pediatric age groups is here. *Rambam Maimonides Med J.* 2021. Online first https://www.rmmj.org.il/issues/online-issue/articles-online/1209.
9. Rose K, Tanjinatus O, Grant-Kels JM, et al. Minors and a dawning paradigm shift in "pediatric" drug development. *J Clin Pharmacol.* 2020;61(6):736–739. https://accp1.onlinelibrary.wiley.com/doi/epdf/10.1002/jcph.1806.
10. Rose K. *Considering the Patient in Pediatric Drug Development. How Good Intentions Turned Into Harm.* London: Elsevier; 2020. https://www.elsevier.com/books/considering-the-patient-in-pediatric-drug-development/rose/978-0-12-823888-2. https://www.sciencedirect.com/book/9780128238882/considering-the-patient-in-pediatric-drug-development.
11. Rose K, Neubauer D, Grant-Kels JM. Rational use of medicine in children—the conflict of interests story. A review. *Rambam Maimonides Med J.* 2019;10(3), e0018. Review https://doi.org/10.5041/RMMJ.10371. https://www.rmmj.org.il/userimages/928/2/PublishFiles/953Article.pdf.
12. Wikipedia. *COVID-19 Pandemic in Israel.* https://en.wikipedia.org/wiki/COVID-19_pandemic_in_Israel.
13. Wikipedia. *COVID-19 Vaccination in Israel.*
14. *Following Israel's Example: The FDA Recommends the Third Vaccine.* Israel Ministry of Health, Press Release; September 18, 2021. https://www.gov.il/en/departments/news/18092021-01.
15. Halpern O. *Israel Struggles With COVID Surge Despite Mass Vaccinations. Israelis Flouting Mask Requirements May Have Been a Main Contributor to the Rapid Spread of the Delta Variant in Israel, Experts Say.* Aljazeera; August 23, 2021. https://www.aljazeera.com/news/2021/8/23/israel-struggles-to-cope-with-surge-of-covid-infections-despite-v.
16. Winer S. *Health Ministry chief says coronavirus spread reaching record heights. As over 10,000 new cases are diagnosed, Nachman Ash tells lawmakers he had hoped recent downward trend would continue.* The Times of Israel; September 14, 2021. https://www.timesofisrael.com/health-ministry-chief-says-coronavirus-spread-reaching-record-heights/.
17. Osbourne H. *Travis County reports first child to die from COVID-19 complications amid rise in pediatric cases.* Austin American-Statesman; August 31, 2021. https://eu.statesman.com/story/news/2021/08/31/austin-child-dies-covid-complications-travis-county-hospital-cases/5663086001/.

References   **229**

18. A. Rodriguez. *Are Schools Contributing to a Spike in COVID-19 Cases Among Kids? Partly, Experts Explain.* USA Today, https://eu.usatoday.com/story/news/health/2021/08/26/covid-child-cases-rising-school-year-stumbles-outbreaks/5573057001/.
19. Edwards E. *5-year-old is first child death from COVID-19-related inflammatory syndrome reported in U.S. Nearly 100 children in the U.S. have been diagnosed with the newly identified syndrome associated with COVID-19.* ABC News; May 8, 2020. https://www.nbcnews.com/health/kids-health/boy-5-dies-covid-19-linked-inflammatory-syndrome-n1203076.
20. *Galveston County mom says 4-year-old got fever in middle of the night, died from COVID just hours later. Health officials said other members of the family had COVID and they don't think the little girl contracted it at school.* KHOU 11 Staff; September 9, 2021. https://www.khou.com/article/news/health/coronavirus/youngest-covid-death-galveston-county/285-d6e954d3-6d85-4548-8d07-6fc229bf4f51.
21. Ray S. *Moderna CEO Says Pandemic Could Be Over Next Year as Vaccine Production Ramps Up.* Forbes; September 23, 2021. https://www.forbes.com/sites/siladityaray/2021/09/23/moderna-ceo-says-pandemic-could-be-over-next-year-as-vaccine-production-ramps-up/?sh=5ef6cd637b9f.
22. Buntz B. *Moderna CEO Predicts the Pandemic Will Be Over in a Year.* Drug Discovery and Development; September 23, 2021. https://www.drugdiscoverytrends.com/moderna-ceo-predicts-the-pandemic-will-be-over-in-a-year/.
23. Segar M. *Moderna Chief Executive Sees Pandemic Over in a Year—Newspaper.* Reuters, Healthcare & Pharmaceuticals; September 23, 2021. https://www.reuters.com/business/healthcare-pharmaceuticals/moderna-chief-executive-sees-pandemic-over-year-newspaper-2021-09-23/.
24. International Committee of Medical Journal Editors (ICMJE). *Recommendations for the Conduct, Reporting, Editing, and Publication of Scholarly work in Medical Journals.* http://www.icmje.org/recommendations/.
25. WHO. *International Classification of Diseases (ICD).* https://www.who.int/standards/classifications/classification-of-diseases.
26. Wikipedia. *International classification of diseases (ICD).* https://en.wikipedia.org/wiki/International_Classification_of_Diseases.
27. US CDC. *International Classification of Diseases, Tenth Revision (ICD-10).* https://www.cdc.gov/nchs/icd/icd10.htm.

Index

Note: Page numbers followed by *f* indicate figures and *t* indicate tables.

A
Access to COVID-19 Tools Accelerator (ACT-C), 85–86
Acute lymphoblastic leukemia (ALL), 29–30
Additional dimensions, 167–168
Agreement on Trade-Related Aspects of Intellectual Property Rights (TRIPS), 184
Agriculture, 20–21
American Academy of Pediatrics (AAP), 227
American Medical Association (AMA), 227
American Psychiatric Association (APA), 142
Andrew Wakefield, 162–163
Animal domestication, 20–21
Antiviral drugs, 96–97
AstraZeneca, 52
Autism and measles, 162–163

B
Bay of Pigs, 111
Bhopal, 121
Big Pharma, 52
Bill & Melinda Gates Foundation, 85–86
Biomedical Advanced Research and Development Authority (BARDA), 4–5, 48
BioNTech, 52
Blockbuster, 31
Bodily, 2–3
Boeing 737 MAX crashes, 113
Bovine spongiform encephalopathy (BSE), 123–124
British Medical Journal (BMJ), 152
BSE, Jacob-Creutzfeld-disease, 123–125
Bushmeat, 19–20

C
Capitalism, 178
Center for Adults with Pediatric Rheumatic Illness (CAPRI), 9–10
Centers for Disease Control and Prevention (CDC), 4–5, 48, 217
Charity, 173
Chimeric antigen receptors (CARs), 29–30
Chinese Center for Disease Control and Prevention (CDC), 73
Chinese vaccines, 73–75
Chronic lymphocytic leukemia (CLL), 29–30
Climate change, global warming, 197–201
Club of Rome, 206
Coalition for Epidemic Preparedness Innovations (CEPI), 85–86
Cold viruses (adenoviruses rAd26 and rAd5), 71–72
Communication, 189–190
Consumer News and Business Channel (CNBC), 73–74
COVAX, 217
COVID-19, 1, 35–36, 56–59, 71–72, 153
   high-tech challenge, 208–209
   internet, 193–195
   lockdowns, 36
   multisystem inflammatory syndrome (MIS), 58–59
   pandemic, 36
   vaccination,age, 57–58, 57–58*t*
   variants, 60
COVID-19 vaccines global access (COVAX), 87
   booster shots, 87
   introduction, 85
   key international organizations, 85–86
   preliminary assessment, 88
Craig Venter, 134
Culture, 24–25, 190
Curevac, 52
Cyber-sphere, 189
Cytokine storm, 42

D

David A. Kessler, 48
Deoxyribonucleic acid (DNA), 36–41
  biosynthesis, 39–40
  nucleotides, 36–37
Department of Health and Human Services (DHHS), 4–5
Developing countries, COVID-10 pandemic, 181–183
Diagnostic and Statistical Manual of Mental Disorders (DSM), 142
Dictatorship of the proletariat, 213–214
Diesel software fraud scandal, 137–138
DNA1. *See* Deoxyribonucleic acid (DNA)
Drugs development, vaccines, 29–30

E

Efficacy of vaccinations, 73, 164–165
Eight drugs, Covid 19, 3
EMA. *See* European Medicines Agency (EMA)
Emergency use authorization (EUA), 2–3
EPA. *See* Environmental Protection Agency (EPA)
EST. *See* Expressed sequence tags (EST)
Ethics committees (ECs), 227
EU Lisbon Agenda, 200
European Centre for Disease Prevention and Control (ECDC), 81
European Health Emergency Preparedness and Response Authority (HERA), 82, 203
European Medicines Agency (EMA), 2–3, 31, 71–72, 81, 193
European Space Agency (ESA), 149–150
European Union (EU), 94
  details, 80–81
  future plans, 81–82
  pandemic in Europe, 79
  preliminary assessment, 83
Expressed sequence tags (EST), 136

F

Fairy tales, oral tradition, radio and television, internet, 191–192
Fake news, 162
FDA emergency use authorization, 2–3
Food and Agriculture Organization (FAO), 206
Free world, 161–162
Fukushima Nuclear plant meltdown, Japan, 114–115

G

GAVI Access to COVID-19 Tools Accelerator (ACT-C), 85–86
Global Times, 74–75
Global warming, COVID-19 pandemic, 201–205
Gross domestic product (GDP), 213–214
Groupthink, 110–111

H

Haemophilus influenzae, 135–136
HERA. *See* European Health Emergency Preparedness and Response Authority (HERA)
Historical epidemics, 1
HIV/AIDS, 184
Human Genome Project (HGP), 133–134
Human immunodeficiency virus/HIV-1 and HIV-2, 19–20
Humanity, communication, 189–191
Hypocrisy, 25–26

I

ICD. *See* International classification of diseases (ICD)
ICMJE. *See* The International Committee of Medical Journal Editors (ICMJE)
Ideologies, politics, abolish poverty, 177–181
Incest, 24–25
Independent panel for pandemic preparedness response (IPPR) report, 101–103, 101t
Influenza (flu), 44
Infodemic, 194
The Institute for Genomic Research (TIGR, now part of the J Craig Venter Institute), 135–136
Institutional Review Boards (IRBs), 227
Intellectual property rights, 183–184
Intergovernmental Panel on Climate Change (IPCC), 198–199
International Classification of Diseases (ICD), 91, 227
The International Committee of Medical Journal Editors (ICMJE), 227
International Health regulations (IHR) and PHEICs, 92–93, 92t
Internet, 189
IPCC. *See* Intergovernmental Panel on Climate Change (IPCC)
Israel, 167

J
The journal "Nature,", 149

K
Kehoe principle, 118
King of Babylon, 213

L
Lead intoxication, 115–116
Lead poisoning, 115–119
Legally, 2–3
Lenin, 179
Letterpress, 191
Life expectancy, 24
Li Wenliang, 41–42
Love Canal/Blackcreek villag, 119–120

M
Mad cow disease, 110
Malthus, 205–206
Manhattan Project, 31
Measles, mumps and rubella (MMR) immunization, 162–163
Medical exemptions, 159
Medical-industrial complex, 151–153
Melancholy, 145
Mercury poisoning, 120–121
Messenger RNA (mRNA), 30
Methylmercury (MeHg), 120
Misconception, weakening of intellectual property, 183–185
Moderna vaccine, 2–3, 52, 226–227
Modern society, public health, 43–45
  Global Alliance for Vaccination and Immunization (GAVI Alliance), 44
  Global influenza Surveillance and Response System (GISRS), 44–45
  Global initiative on sharing avian influenza data" (GISAID), 44–45
Moncef Slaoui, 48
Multiple in flammatory syndromes (MIS) in children (MIS-C), 9

N
National Aeronautics and Space Administration (NASA), 111–112
National Human Genome Research Institute (NHGRI), 136
National Institutes of Health (NIH), 4–5, 48, 133–134
National Library of Medicine (NLM), 142–143
National Socialism, 179
"Nature," European Union (EU) science budget deliberations, 2019, 149–151
Nebuchadnezzar, 213
New England Journal of Medicine (NEJM), 151–152
Noble savage, 21
Non-medical exemptions, 159
North America Space Agency (NASA), 197–198

O
Office of Scientific Research and Development (OSRD), 31
Olympic Athletes from Russia (OAR), 75
Operation Warp Speed (OWS), 4–5, 183, 216
Organization for Economic Co-operation and Development (OECD), 143
Our Planet, Our Future, 202

P
Pandemic COVID-19, 1–13
Paradigm shifts in science, 141–142
Pediatric studies, 225
Penicillin, 30
Pfizer/BioNTech vaccine, 6, 71–72
Pharmaceutical industry, 180
Pharmaceuticals and Medical Devices Agency (PMDA), 31
Philanthropy, 173–174
Placebo/I will please, 31–32
Police control, 76
Police state, 217
Politics, websites, real world, 192–193
Polyvinyl chloride (PVC), 119–120
Potëmkin villages, 213–214
Prevention, Traditional avenues, 45–47
Predicted effect, climate change, humans, 201
Privileges, youth, 208
Promises to abolish poverty, 180
Public health, 7, 216
Purchasing power parity (PPP), 214–215

R
Receptor antagonists/substances, 30–31
Reclaim Economics, 207
Religion, 22–23
Remdesivir, 3
Rhetoric, 182
Russian vaccine, 71–73

S

Science and society, 144–147
Science of science, 146–147
Scientific warnings, 205–208
Serotonin reuptake inhibitor (SSRI), 31
Severe acute respiratory syndrome coronavirus 2 (SARS CoV-2), 41–43, 61–62, 137, 164–165, 213
   acute respiratory distress syndrome (ARDS), 42
   angiotensin-converting enzyme 2 (ACE2), 41
   fatal infections, 41
   middle east respiratory syndrome (MERS), 41
Shotgun sequencing, 135
Simian immunodeficiency virus (SIV), 19–20
Sinopharm, 73
Sinovac vaccine (CoronaVac), 73
Social framework, responsibility, 174–176
Social inequality, 176–177
Sociology, 176
Social function of science, 146–147
Special edition, 150 years, "Nature," 2019, 147–149, 147t
Special Program for Research and Training in Tropical Diseases (TDR), 98–99
Spiritual world, 189–190
Stéphane Bancel, 226–227
Sustainable Development Goals (SDGs), 204

T

Terrible important professors (TIPs), 13, 137, 216
Tetraethyl lead (TEL), 141–142
Thalidomide, 31
TIGR assembler, 135–136
Tobacco smoking, 121–123
Tokyo Electric Power Company (TEPCO), 114–115
Trade-Related Aspects of Intellectual Property Rights (TRIPS), 97–98
Transparency of clinical data, 74, 216
Treatment, 53–56, 55t
TRIPS. *See* Agreement on Trade-Related Aspects of Intellectual Property Rights (TRIPS)
Trust, 216

U

United Nations Environment Programme (UNEP), 198–199
United Nations Framework Convention on Climate Change (UNFCCC), 199
UN sustainable development goals (SDG), 147
US Environmental Protection Agency (EPA), 137–138
US Food and Drug Administration, 7–9
US space shuttle disasters, 111–113

V

Vaccines, 47–53
   alliance, 86
   history, 3–4, 4f
   nationalism, 87
   Operation Warp Speed (OWS), 48
   Virus neutralizing antibodies (VNAs), 50
Vaccines Global Access (COVAX), 72
Variant Creutzfeldt–Jakob disease (vCJD), 123–124
Vitro molecular testing, 53

W

Walvax Biotechnology, 75
War, 94, 219–220
Warfare, 21–22
Warp drive, 4–5
White knights, 166–167
WHO basics, Public Health Emergencies of International Concern (PHEICs), 91–92
WHO IPPR recommendations, Assessment, 103–104
WHO IPPR report, 104–105
Whole-genome random sequencing, 135–136
WHO's life, 93–96
Wildfires, 198
World Health Organization (WHO), 36, 71–72, 99–101, 193, 217
   Progress in healthcare, 96–99
World Meteorological Organization (WMO), 198–199
World Trade Organization (WTO), 184

Printed in the United States
by Baker & Taylor Publisher Services